Order in the Universe
by
Henry Ohlef

Copyright © 2021, all rights reserved

While this non-fiction book is a work of fact and opinion, in general the names, characters, places or incidents have been documented by previous authors, publishers, or opinions of others. Any resemblance to actual people of today or within the author's lifetime or events or locales that might be concrued as referencing their actions or events due to some similarity is entirely coincidental.

Printed by Kindle Direct Publishing, an Amazon.com company.

Published in the United States by Temporal Publishing, an enterprise created by Henry L. Ohlef.

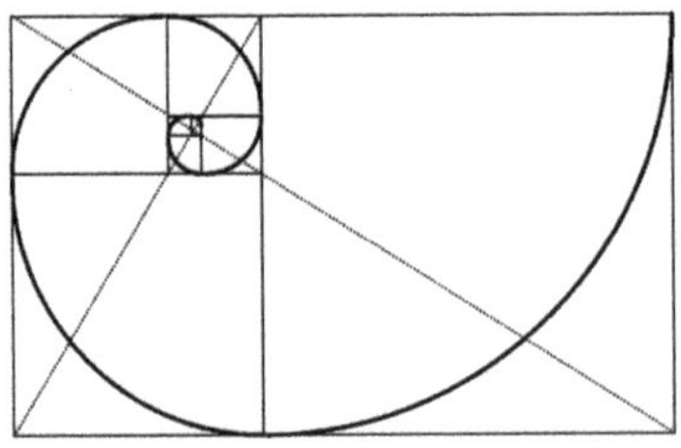

Illustrations: All illustrations in this book are attributed to their respective sources in Appendix B.

Front Cover: Universe graphic, ID 175872873 © Mariia Vasileva | Dreamstime.com
Back Cover: Man Silhouette, ID 161060914 © Alexandra Barbu | Dreamstime.com
Back Cover: Mud Puddle, ID 191402723 © Coryn9 | Dreamstime.com
Back Cover: Saturn Wallpaper, ID 192710190 | © Elen33 |Dreamstime.com

Preface

As an overview, I would say that there are three categories of human understanding of science: The theoretical scientist, the applied technician, and the public. As an electrical engineer, I would say that I belong to the class of applied technician. I will never understand fully the theoretical scientist and I am wondering whose fault that is. More importantly, do theoretical scientists see the forest through the trees? At the heart of this book is the question of the forest.

After listening to a theoretical lecture on Einstein's general theory of relativity for four hours where gaps in the derivation were assumed as too difficult or inconvenient to present, the question becomes whether or not internal knowledge leads to the protection of theories that no one else understands. It becomes a self-fulfilling prophesy. Hidden knowledge protects the professional but does not illuminate the public.

Anyone with a science or mathematics background will retain an interest in science their whole lives. Anyone with an interest in the movies can barely miss being inundated with scientific myths. Is there such a thing as political science, the search for the truth?

All my adult life I have had a friend that talked science with me. Many of his witticisms and understandings are in this book. However with his personal past, he has had no time to collaborate on this book. I owe him my thanks for these many years of friendship.

There are over 100 illustrations in this book. Who knew that it would be so hard to present public material about the root of knowledge without bumping into copyright restrictions? Does Johannes Kepler's 1619 book <u>Harmonices</u>

<u>Mundi</u> require a royalty fee to reproduce his illustrations? I thought that knowledge was free, like water and air. Apparently, I was wrong. However, there are some websites that I visited where knowledge was mostly free: Wikipedia, NASA, and Creative Commons. I thank them for their service. For the rest, the attributes required are listed in Appendix B.

If you have no memory and you step into a mud puddle, then you are likely to do it again. The relevance of this book depends on whether or not our Universe is destroying itself or creating new life. There is plenty of scientific pursuit of the former and not enough emphasis on the latter. The Universe was not built on the flip of a coin. By understanding the "Order in the Universe," a guide becomes evident to help create new life and thus new success. This success may seem lost in the stars, but it is very much with every human being today. Success or failure, the curious mind will want to know what our purpose is within the laws of Nature.

This new Century is libel to burst at the seams. My science fiction book, <u>The Second Judgment</u>, is a fanciful story of love and morality during World War Three. As technology advances, and access to it becomes harder or more costly to find, it is important to understand the issues in any field. Our world must not be buffaloed into taking someone's word for knowledge. We are at a critical Fibonacci series juncture in the history of the United States, 89 years after the great depression of 1929 and beyond. I believe that knowledge should be investigated and challenged. As an industrial teacher, I have spent many hours in the classroom. Anyone who has genuine knowledge of a discipline will welcome the challenge and thank you for it.

My wish is for better science within a better world. I hope that you find a truth in this writing that is harmonious with your thoughts.

Henry Ohlef --- December 1, 2020

List of Illustrations

Table of Contents

Chapter 1: The Great 20th Century Debate

In the spring of 1926 just outside the University of Berlin, two of the most famous scientists of the 20th Century went for a walk. Werner Heisenberg had just delivered a mathematical overview lecture on the new science of Quantum Mechanics to a select group of physicists that developed the science. In his autobiography, Werner Heisenberg (WH) remembered his conversation with Albert Einstein (AE). The full text of his remembrance is reproduced in Appendix A. What follows is an abridged version that occasionally is paraphrased for clarity:

AE   What you have just told us sounds very strange. You assume the existence of electrons inside the atom .... but you refuse to consider their orbits.

WH We cannot observe electron orbits inside the atom, but the radiation which the atom emits during discharges enables us to deduce the frequencies and corresponding amplitudes of its electrons.

AE   But you don't seriously believe that none but observable attitudes must go into a physical theory?

WH Isn't that precisely what you have done with relativity? After all you did stress that in fact it is impermissible to speak of absolute time simply because absolute time cannot be observed.

AE   Possibly I did use this kind of reasoning, but it is nonsense all the same. .... it is quite wrong to try founding a theory on observable magnitudes alone. In reality, the very opposite happens. It is the theory that decides what we can observe.

WH You must appreciate that observation is a very complicated process. .... we must be able to tell how nature functions, must know the natural laws at least in practical terms, before we can claim to have observed anything at all. .... we nevertheless assume that existing laws function in such a way that we can rely upon them and hence speak of 'observations'.

AE   And in your theory, you quite obviously assume that the whole mechanism of light transmission from the vibrating atom to the spectroscope or to the eye works just as one always supposed it does, that is, according to Maxwell's laws. If that were no longer the case, you could not possibly observe any magnitudes you call observable. You claim that you are introducing none but observable magnitudes is therefore an assumption .... [that may clash] with the old description of radiation phenomena in the essential points. You may be right, of course, but you cannot be certain.

WH ..... thought economy surely goes back to Mach ..... Even when small children learn to speak and to think, they do so by recognizing the possibility of describing highly complicated but somehow related sense impressions with a single word, for instance, the word "ball".

AE   All this sounds very reasonable, but we must nevertheless ask ourselves in what sense the principle of mental economy is being applied here. When the child forms the concept 'ball', does he introduce a purely psychological complication, or does this ball really exist? .... you will also have to say, in a cloud chamber we can observe the path of the electrons. At the same time, you claim that there are no electron paths inside the atom. This is obvious nonsense, for you cannot possibly get rid of the path simply by restricting the space in which the electron moves.

WH For the time being, we have no idea in what language we must speak about the processes inside the atom. .... Hence it is probably much too early to solve the difficulties you have mentioned.

AE Very well, I can accept that. Quantum Theory as you have expounded it in your lecture has two distinct faces. On the one hand, as Bohr himself has rightly stressed, it explains the stability of the atom; it causes the same forms to reappear time and again. On the other hand, it explains that strange discontinuity or inconstancy of nature which we observe quite clearly when we watch flashes of light on a scintillation screen. In your quantum mechanics, you will have to take both into account. .... As you know, I suggested that when an atom drops suddenly from one stationary energy value to the next, it emits the energy difference as an energy packet, a so-called light quantum. In that case, we have a particularly clear example of discontinuity. Do you think that my conception is correct?

WH Bohr has taught me that one cannot describe this process by means of traditional concepts; i.e., as a process in time and space. Whether or not I should believe in light quanta, I cannot say at this stage.

At this point, knowing the history of 20th Century physics helps to unpack some of the issues raised in this dialog. Let us try a few:

- "Frequencies and corresponding amplitudes": The description of electrons exhibiting wave properties means that electrons have orbits and that they have quantum energy states with amplitudes depending on the orbit position around the nucleus.
- "Observations": The observation of a "ball" depends on whether you can see the ball or whether at the quantum level, the ball disappears inside a "cloud chamber". In other words at the quantum level, the position of an electron is unknown, only its wave properties can be measured.
- "Time and space": The distance between observing the "ball" and having the "ball" disappear symbolizes the difference between Newtonian mechanics and Quantum Mechanics. The difference "sounds very strange". Bluntly put, Einstein wants theorems of physics to be observable in "time and space". Heisenberg will state that quantum observations are uncertain.
- "Mental economy": This phrase does not sound like a scientific term, but it highlights that observations made in "time and space" do not match frequency and amplitude observations that are independent of time and, therefore, independent of position. Simply put, what is reality? Are we living in a world of sight, touch, and smell or are we living in an alternate reality?
- "Discontinuity": If, for example, an electron in orbit around the nucleus of an atom gets hit by a photon of light, the energy absorbed allows the electron to make a discrete quantum energy jump to a higher and more distant orbit. As long as the electron is in orbit, its continuous wave properties can be analyzed with the mathematics of calculus. However, if the electron jumps to a new orbit, then this is an example of a discontinuity and must be treated with discrete difference mathematics.

The debate is just getting started. From 1927 to 1935, the top theoretical scientists in Europe met each year in Brussels for the Solvay Congress. The Solvay debates may be the most famous debates in science history to date. Einstein did not attend every conference.

In 1927, Einstein stood up and objected after Bohr's informal presentation of his complementarity principle. In addition, the Max Born presentation stated that he and others considered quantum mechanics to be a closed theory, whose fundamental physical and mathematical assumptions are no longer susceptible of any modification. Einstein's general reply, "Quantum mechanics is very impressive. But an inner voice tells me that it is not yet the real thing. The theory produces a good deal but hardly brings us closer to the secret of the Old One."

For the next few days, Born and Einstein met and argued. Einstein made the following argument: "Consider a particle passing through a slit of width d. The slit introduces uncertainty in momentum approximately because the particle passes through the wall. But, let us determine the momentum of the particle by measuring the recoil of the wall. In doing so, we find the momentum of the particle to arbitrary accuracy by conservation of momentum." This question directly attacked Heisenberg's new Uncertainty Principle; namely, there is a point of quantum measurement that is so small that it cannot be measured.

Neils Bohr responded to the premise. You must measure the recoil [resistance] of the wall to the same accuracy as the momentum of the photon. Since by Heisenberg's Uncertainty Principle, the momentum is uncertain, then the wall recoil is uncertain. For the quantum community, Bohr's reasoning had won the day.

In 1930 Einstein came prepared. "Consider a box containing electromagnetic radiation and a clock which controls the opening of a shutter which operated so quickly that it would allow only one photon to escape at a time. The box would first be weighed exactly. Then, at a precise moment, the shutter would open, allowing a photon to escape. The box would then be weighed again."

Einstein's idea was that by his own formula, energy equals mass times the speed of light squared, because the mass of the photon was known, and the speed of light was known, then the energy

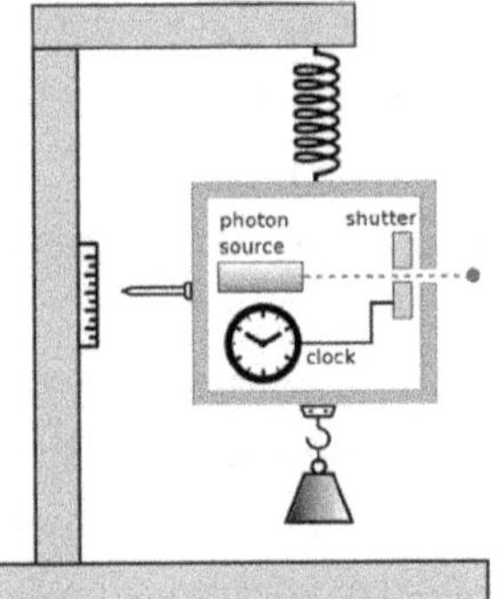

Figure 1: Einstein Photon Box

loss would account for the new weight of the box. Neils Bohr spent a sleepless night. Simply put, Bohr's response was that after emitting the photon, by the red shift in light frequency caused by the speed of light, the clock inside the box has changed the frequency of time itself. Therefore to the observer, the time of emission from the box is uncertain, and therefore by the Uncertainty Principle, the momentum (or energy expended) is uncertain.

Worn down by the debate about the uncertainty of position, Einstein co-authored a paper about particle entanglement in 1935 that placed the quantum community in a position of having to defend the completeness of the theory. "Consider two particles that have collided or which have been created in such a way that they have properties which are correlated. The total wave function for the pair links the positions of the particles as well as their linear momenta."

Neils Bohr was stunned by this argument. At the time he acknowledged that there may be mechanical problems that need to be considered. He made the weak argument that the two particles are one system and therefore one quantum function (not two) that must abide by the Uncertainty Principle. However by 1935, the general theory of quantum physics was well established and did not need to be defended. Consider the following Nobel Prize for physics laureates:

Year	Laureate	Contribution to Physics
1918	Max Planck	Discovery of energy quanta
1921	Albert Einstein	Discovery of the law of the photoelectric effect
1922	Neils Bohr	Investigation of the structure of atoms and of the radiation emanating from them
1929	Louis de Broglie	Discovery of the wave nature of electrons
1932	Werner Heisenberg	Creation of quantum mechanics, the application of which has, inter alia, led to the discovery of the allotropic forms of hydrogen
1933	Erwin Schrodinger Paul Dirac	Discovery of new productive forms of atomic theory (electrons, protons, neutrons)

Einstein and Bohr continued to write each other. In a letter to Bohr, Einstein wrote what was to become the most famous scientific statement of the 20th Century, namely, "I am convinced, in any case, that He [God] does not play with dice."

Based on these debates, the search for the truth seems somewhat esoteric. Daily lives were not much affected by the Solvay Congress. Yet in 1942, Werner Heisenberg as a card carrying member of the Nazi party was asked to meet with Albert Speer, the head of the SS secret police for Adolph Hitler. He was asked what it would take to make a nuclear bomb. Heisenberg replied that the potential human resources would require 1200 men and much time. His understanding of the response was that Hitler could not spare the human resources. Besides, if Heisenberg were in charge, they might produce nuclear fission material, but they would be nowhere close to a bomb.

Albert Einstein, a pacifist and a Jew, left Germany in 1935 and eventually became a United States citizen in 1940. He wrote to President Franklin Roosevelt in 1939, that German scientists had the capability of building a nuclear bomb. This impetus spurred Roosevelt to begin the Manhattan Project and develop the atomic bomb.

The aftermath of these debates caused the most famous living scientists to be on the opposite side of the world fence. One was on the side of saving the world and the other was on the side of destroying the world. Besides their origins, could it be possible that their different humanitarian views were affected by their science and ultimately by the debates?

For example, physics in "time and space" generally means solving a linear equation and receiving both position and momentum information. Einstein's famous equation for energy provides these answers:

$$(1) \qquad E = mc^2 \qquad where$$

- $E = kinetic\ Energy$
- $m = relativistic\ mass$
- $c = speed\ of\ light\ constant = 186,000\ miles\ per\ second$

In the equation, the increased relativistic mass (m) of a body times the speed of light (c) squared is equal to the kinetic energy (E) of that body. The equation unlocked the secret to the atomic bomb but also stands for abundant energy that when harnessed could feed the world for millenniums to come.

Heisenberg's Uncertainty Principle is a probability equation that places limits on what can be quantumly solved in physics. To be "observed", position times momentum must be greater than Plank's constant:

$$(2) \qquad \sigma x * \sigma p = \frac{h}{2\pi} \qquad where$$

- $\sigma x = position$
- $\sigma p = momentum$
- $h = Plank's\ constant = 6.62607015 \times 10 - 34\ joules - second$

The formula beyond a certain point means that quantum experimental results will be uncertain. Position and velocity or waveform amplitude cannot exist at the same time.

In 1935, Erwin Schrodinger in response to Einstein's entanglement paper wrote a paper with Quantum Mechanics most colorful invention – Schrodinger's cat:

SCHRÖDINGER'S CAT

Figure 2: Schrodinger's Cat

If entangled particles were super-positioned so that its nucleus was locked in a box with a cat, then if the nucleus decayed by emitting a light photon, the cat would die; otherwise, the cat would live. While scientists waited for results, whether or not the cat is alive or dead was uncertain. Ironically, Einstein in his breakthrough year of 1905, the year of relativity, also formulated the Brownian motion of a particle. If a particle from an origin moves randomly in any direction, then its position is unknown but it is bounded by the number of steps (n) times the square root of 2 ($\sqrt{2}$). By examining the cat image above, It is clear that the image indicates random behavior. As the seconds tick by, the uncertainty changes in a random fashion. If a particle wave heads in one direction and after a time travels in the opposite direction, sooner or later, the waves cancel each other. Random behavior is chaotic behavior. The result is the dust of destruction. There is no order.

We have reached the crux of the matter. Does God roll the dice? If so, there is no order anywhere. The Universe will destroy itself. On the other hand, if we can find order, then the Universe and its inhabitants will flourish in abundance. This is our hypothesis; this is our quest.

Chapter 2: Third Order Polynomials

Ancient Geometry and its champions are easily ignored in today's educational world. The sharp minds of Plato and Pythagoras are hard to understand, but their mathematical theories have much use in our Century. They solved nonlinear problems that somehow have eluded our modern times. From his Timaeus, Plato builds a model of the Universe out of elementary building blocks:

"In the first place it is clear to everyone that fire, earth, water and air are bodies, and all bodies are solids. All solids again are bounded by surfaces, and all rectilinear surfaces are composed of triangles. There are two basic types of triangles, each having one right angle and two acute angles; in one of them these angles are both half right angles, being subtended by equal sides, in the other they are unequal, being subtended by unequal sides. This we postulate as the origin of fire and the other bodies, our argument combining likelihood and necessity; their ultimate origins are known to god and to men whom god loves. We must proceed to inquire what are the four most perfect possible bodies which, though unlike one another, are some of them capable of transformation into each other on resolution. If we can find the truth about the origin of earth and fire and the two mean terms; for we will never admit that there are more perfect visible bodies than these, each in its type. So we must do our best to construct four types of perfect body and maintain that we have grasped their nature sufficiently for our purpose. Of the two basic triangles, then, the isosceles has only one variety, the scalene an infinite number. We must therefore choose, if we are to start according to our own principles, the most perfect of this infinite number."

The square or cube is the most perfect Platonian solid. However, there were no irrational numbers in Plato's time. To slice a square was a sacred cut. The triangle and the tetrahedron are the strongest of the physical bodies.

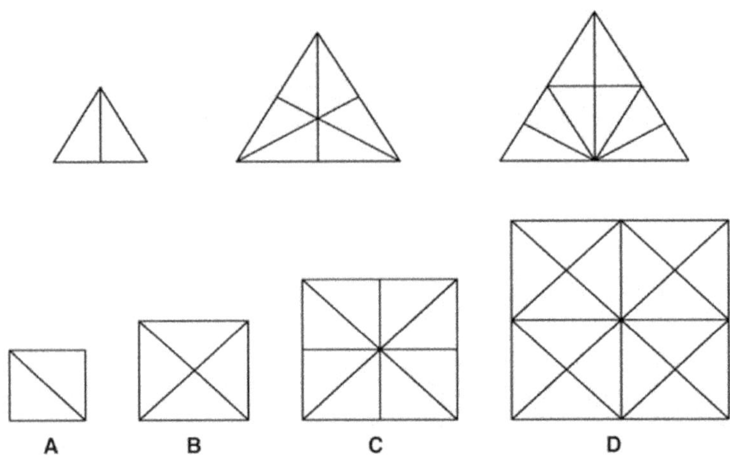

Figure 3: Plato geometry

Again from Timaeus:

"First will come that form which is primary and has the smallest components, and the element thereof is that triangle which has its hypotenuse twice as long as its lesser side. And 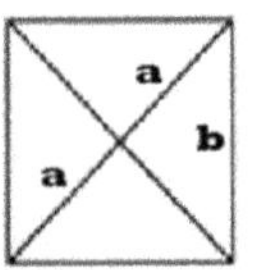 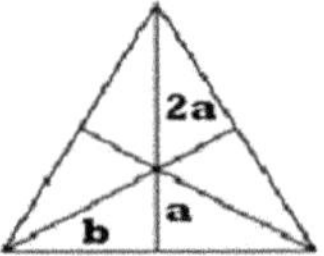

Figure 4: Plato Proportions

when a pair of such triangles are joined along the line of the hypotenuse, and this done thrice, by drawing the hypotenuses and the short sides together as to a centre, there is produced from those triangles, six in number, one equilateral triangle."

From the square design, the: dimensions of the triangle are proportioned as follows:

$$(3) \qquad \frac{a}{b} = \frac{b}{2a}$$

This example is a three term proportion, 'a' is to 'b' as 'b' is to 'c', where 'b' is the "observer" between 'a' and 'c'. A three term proportion is the basis of the proper observer in the scientific method. Nothing is proved, but the observation can be used to form a scientific hypothesis. Given that 'a' equals 1, then :

$$(4) \qquad a = 1, \quad b = \sqrt{2}, \quad c = 2a = 2$$

Solving for the roots of a third order polynomial may shed some light on these proportions:

$$(5) \qquad ax^3 + bx^2 + cx + d = 0 \quad or$$

$$(6) \qquad x^3 + \frac{b}{a}x^2 + \frac{c}{a}x + \frac{d}{a} = 0 \quad or$$

$$(7) \qquad (x - \alpha)(x - \beta)(x - \gamma) = 0 \quad where \ \alpha, \beta, \gamma \ are \ the \ 3 \ roots$$

Multiplying the root equation, gathering terms, and equating them with our polynomial coefficients leads to the following results from Vieta's formula for any polynomial, no matter the order:

$$(8) \qquad \alpha + \beta + \delta = -\frac{b}{a}$$

$$(9) \qquad \alpha\beta + \alpha\gamma + \beta\gamma = \frac{c}{a}$$

$$(10) \qquad \alpha\beta\gamma = -\frac{d}{a}$$

Hopefully for these three coefficients of a third order polynomial, you can see the most significant ancient ratios in mathematics; namely, the arithmetic mean (AM), the harmonic mean (HM) and the geometric mean (GM). With these means, the ancients solved nonlinear problems:

$$(11) \qquad HM = \frac{GM}{AM} \qquad or \qquad GM = HM * AM$$

There are many clever ways to solve polynomial equations. Among the early tries, using plus or minus 1 and adding the coefficients works for many classroom problems. Another way simply is to use the raw third order formula for equations to the third power.

(12)

$$x = \sqrt[3]{\left(\frac{-b^3}{27a^3} + \frac{bc}{6a^2} - \frac{d}{2a}\right) + \sqrt{\left(\frac{-b^3}{27a^3} + \frac{bc}{6a^2} - \frac{d}{2a}\right)^2 + \left(\frac{c}{3a} - \frac{b^2}{9a^2}\right)^3}}$$

$$+ \sqrt[3]{\left(\frac{-b^3}{27a^3} + \frac{bc}{6a^2} - \frac{d}{2a}\right) - \sqrt{\left(\frac{-b^3}{27a^3} + \frac{bc}{6a^2} - \frac{d}{2a}\right)^2 + \left(\frac{c}{3a} - \frac{b^2}{9a^2}\right)^3}} - \frac{b}{3a}.$$

Wow! This formula is no fun. Fortunately, there are software engineers that can program a calculator to solve this equation. Nevertheless, let us solve the following polynomial with the coefficient tools of Equations 6, 7, 8:

$$(13) \qquad x^3 - 4.4142x^2 + 6.2426x - 2.8284 = 0$$

In general, this polynomial looks unusually challenging. Before we solve, let us examine a typical third order polynomial graph with three roots. Note, that a third order polynomial always has two curves and, in this case, intersects the x-axis three times. In general, if none of the coefficients are negative, then the wavy line exists above the x-axis and has no roots.

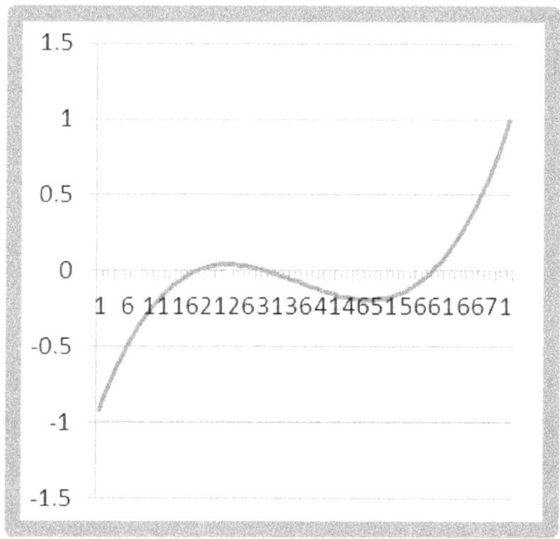

Figure 5: Third Order Polynomial

Luckily, the middle intersection root also happens to be the inflection point. If you rotate the graph 180 degrees around the inflection point, you get the same graph. This allows us to select one of the equations 6, 7 and 8, and will allow us to solve for the first root. If the lower and upper roots are equally distant from the middle root, then equation 6 will solve as:

$$(14) \qquad 3 * AM = -\frac{b}{a}$$

If the root spacing forms a geometric progression, then the geometric mean when multiplied three times will equal the multiplication of the roots α, β, γ. Equation 8 will solve as:

$$(15) \quad GM^3 = -\frac{d}{a}$$

$$(16) \quad \text{Therefore,} \quad \text{Eq. 7} = \text{``HM''} = \frac{GM^3}{3AM} = \frac{\alpha^3}{3\alpha} = \frac{\alpha^2}{3} = \frac{c}{a}$$

Let us try the geometric solution:

$$(17) \quad \alpha^3 = -\frac{d}{a} = 2.8284 \qquad \alpha = \sqrt{2}$$

By root division, the third order polynomial can be reduced to a quadratic equation and easily solved. The roots are $1, \sqrt{2}, 2$ – the same numbers used to construct Plato's isoceles triangle.

The relationship between the roots and the coefficients can be verified:

$$(18) \quad \text{Equation 8} = \alpha + \beta + \gamma = 1 + \sqrt{2} + 2 = -\frac{b}{a} = 4.4142$$

$$(19) \quad \text{Equation 9} = \alpha\beta + \alpha\gamma + \beta\gamma = \sqrt{2} + 2 + \sqrt{2} * 2 = \frac{c}{a} = 6.2426$$

$$(20) \quad \text{Equation 10} = \alpha\beta\gamma = 1 * \sqrt{2} * 2 = -\frac{d}{a} = 2.8284$$

The relationship between ancient mathematics and modern times is not good. Yet there are architecture examples from all over the world that still stand using cut stone blocks transported for miles that no modern forklift could even budge.

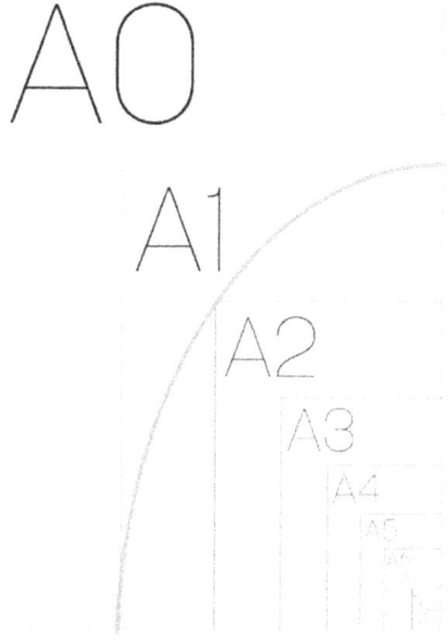

Figure 6: Paper Sizes

Figure 6 is an example of squares within squares and arcs along the diagonals representing the transformation of squares labelled A1, A2, A3, A4. The lettering represents the standard sizes of writing paper for both American and European sizes. Can you believe that a standard piece of A1 paper in America is 8.27 by 11.69 inches? Why has not some mathematician suggested 8 by 11 or 7 by 12 instead of these unusual measurements? One reason is that the largest sheet A0 is 1 meter square in area and the other sheets were derived from this measurement. But the real reason is the "sacred cut" of a square that existed with Plato and his ancestors. This square root of 2 cut along the diagonal and intersecting each arc at the tip of the square transforms each square into a rectangle whose side geometrically grows by the square root of two. These sides in inches respectively are 8.27, 11.69, 16.54, 23.39, and 33.11. Our standard size of writing paper has existed for 5,000 years.

Chapter 3: Light and Time

It is not commonly understood that "time" does not exist. You cannot see time, feel, smell, or touch time. Time is an abstract concept. Yet there are four seasons in most countries: spring, summer, fall, and winter. Abstractly then, time is a useful concept. Apparently, there are 50 atomic clocks around the world that radiate time. They don't always agree because gravity is less in the Himalayas than in Florida. By some standard, a correction of one second was made in 2017.

Figure 7: Apollo and the Sun

The God Apollo, who in his chariot pulls the sun across the sky during the day, has been given a different mission. He is to move the Andromeda Galaxy to a different space in the Universe. While doing so, the time values between the Andromeda Galaxy and the Milky Way have changed. It is as if the space-time of the move was in a parallel universe to the space-time of the Earth. If there were also parallel planets, then when the 41 year old Paul Revere took his famous ride in 1775, Michelangelo may be painting the Sistine Chapel in the Andromedan year of 1512. Because of the faster galaxy, time as experienced on Earth appears slower in the Andromeda Galaxy. With respect to zero A. D., when the second Paul Revere has his ride, the earthly Revere would be 612 years old.

The reason for all this confusion is gravity and the speed of light. Albert Einstein's General Theory of Relativity deals with defining the coordinate systems of space-time coordinates for two objects moving at different velocities. Space-time coordinates are in four dimensions: length, width, height, and time. Four dimensions are difficult to visualize. As a difference in time can be the measure of momentum, the suggestion is that Apollo pulls the dimensions of space over time at a certain velocity; the faster the velocity, the longer the time interval, and, therefore, the slower the time.

In addition to the special coordinates, the famous solar eclipse experiment of 1919 proved that stars could be seen from directly behind the sun due to gravity's ability to curve space. Proof of Einstein's General Theory of Relativity was experimentally obtained. Therefore, Cartesian coordinates will not suffice. The Universe must be modelled as curved space for large bodies with gravity. Spherical coordinates are required whereby the shortest distance is a line that lies on the surface of a sphere. The tensor equations developed by Einstein are complicated.

The difference in clock time for two objects travelling at different velocities is the same for earthly velocities but if one time travels near the speed of light, then its time is dilated:

$$(21) \qquad \Delta T' = \frac{\Delta T}{\sqrt{1-\frac{v^2}{c^2}}} \qquad \text{where}$$

- ΔT = stationary time of observer
- $\Delta T'$ = Time in motion
- v = velocity of observed
- c = speed of light

For velocities much lower than the speed of light, there is no change in the respective time clocks, but as velocity approaches the speed of light, the denominator shrinks and the time in motion grows much bigger. As discussed previously, relativity suggests that Paul Revere on Earth grows older more quickly. If you travel at a constant speed such as a commuter train, you do not feel the speed, but if you decide to take a year's vacation on a spaceship travelling at 25% the speed of light, your younger brother or sister may have great grandchildren by the time you get back.

Between 1880 and 1887, Albert Michelson tried to measure the speed of light. At the time, the speed was not thought of as a constant. Also at the time, the invisible ether was thought of as the medium through which light traveled. Sound travels through air and cannot be heard in a vacuum. The discovery of the ancient ether would be a big scientific breakthrough.

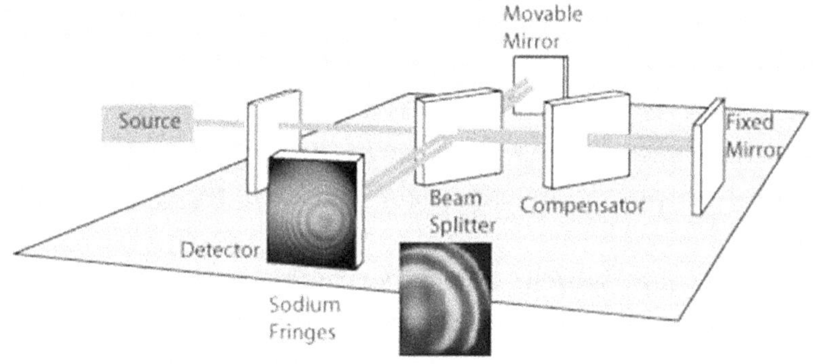

Figure 8: Michelson-Morley Experiment

The famous 1887 experiment goal was to have two beams of light arrive at an inferometer and resolve into frequency pictures of concentric circles. If the two beams are different indicating a time delay, then the center of the picture will be a dark circle (shown as the picture on the right). If the two beams arrive at the same time, then the circle in the center will be filled (as on the picture on the left).

From the beam splitter there are four arms of light. These arms are different entities with different properties. The beam splitter has a transparent part referred to as a slit that allows light to go straight through. However, part of splitter, which is a mirror, is tilted to reflect the beam to the left. Now the split beams hit mirrors perhaps at different times and are reflected back to the center. Potentially, the beams have changed with respect to the initial beam. Now potentially three different beams meet in the center at the beam splitter. From the beam splitter, two beams of light originating from the reflective mirrors arrive at the inferometer (detector). Frankly, what we wind up with is a four term proportion.

$$(22) \qquad \frac{BeamA}{BeamB} \neq \frac{BeamC}{BeamD} \quad \text{or} \quad \frac{BeamA}{BeamC} \neq \frac{BeamB}{BeamD} \quad \text{or} \quad \frac{BeamA}{BeamD} \neq \frac{BeamB}{BeamC}$$

If the times that the two beams are detected are different, which is the positive result of the experiment, then the above inequalities hold. This is not the scientific method. There is no observer or there are two observers. If not eating ice cream to go on a diet is equated with a tree falling in the redwood forest, then you have a four term proportion of unrelated facts. Only a politician can make sense of those facts; not a scientist.

The table on which the experiment took place can be rotated to maximize the effect of the Earth's rotation around the sun. The Earth orbits around the sun at a speed of 67,000 miles per hour. If light is traveling with the Earth's orbit, then the beam should reach its destination slightly slower. On the other hand, if the beam is traveling against the Earth's orbit, then the destination should meet the beam in a faster time. There should be a time difference between the two beams thus demonstrating that light travels through an ether.

Unfortunately, Michelson deemed his experiment a failure. The result for all angles was a bright spot in the middle of the circle pattern (the display on the left). Einstein concluded that the speed of light at 186,000 miles per second was a constant, which leads to the conclusion that there will never be a difference in the interference pattern between two beams of light. When Einstein introduced his Special Theory of Relativity in 1905, he was unaware of the Michelson and Morley experiment. Yet this failed result directly leads to relativistic space-time. Albert Michelson won the Nobel Prize for physics in 1907.

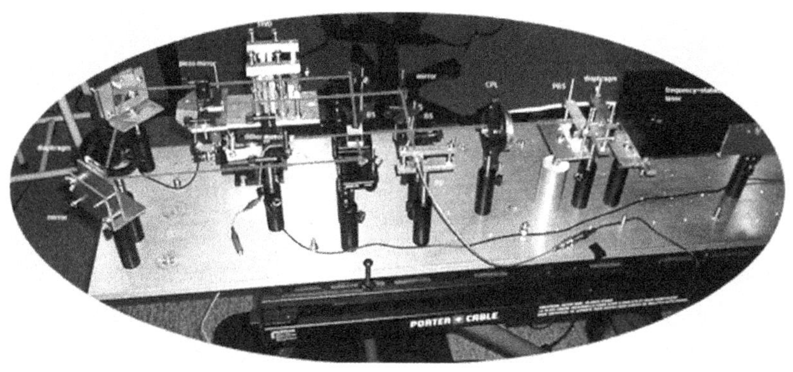

Figure 9: Marett Reconstruction of Silvertooth Experiment

In 2011, Doug Marett tried to reconstruct the first experiment of Earnest Silvertooth in 1989. In turn, Silvertooth was trying to reproduce the Michelson-Morley experiment. Silvertooth, an expert in optics, discovered an interference pattern unlike Michelson, and the results suggested the existence of ether. Unfortunately, Silvertooth's laboratory work was not readily disseminated, and so Marett decided to find out the truth. He built a moving lab table emulating Silvertooth's experiment whereby distances could be adjusted to create standing waves.

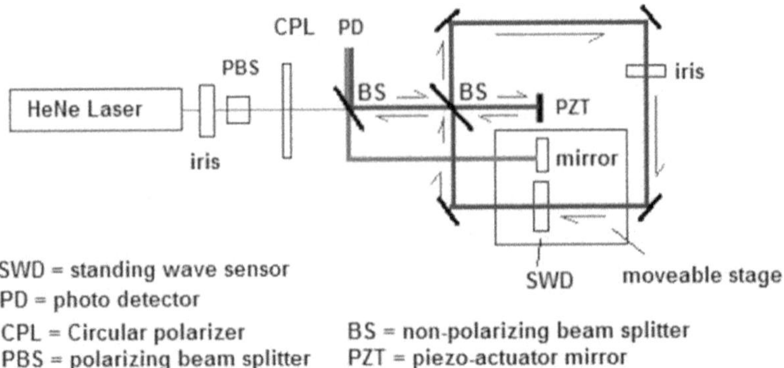

Figure 10: Schematic of Marett's Experimental Table

The busy diagram essentially disguises the two light beams that are to be compared. There is an initial setup. The platform of the standard wave sensor (SWD) must be moved so that the "square" beam is sensed as a frozen electronic wave. Then the inert distance traveled from the SWD mirror and the average beam length from the actuator PZT mirror to the photo diode are the same. Again, the lengths may be the same, but does the speed of the Earth affect the time it takes for both beams of light to travel? One fact that the experiment has in its favor is after initiation or setup, there are two beams of light and photo diodes (PD) to change the light into electronic waveforms. This creates a three term proportion. The oscilloscope results below show the displacement.

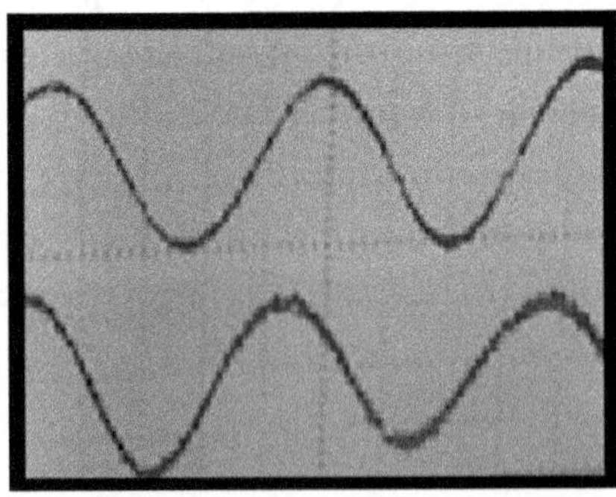

Figure 11: Oscilloscope Results of Marett's Experiment

While this initial result looks conclusive, the table's position had to be aligned with the Earth's space direction to obtain a theoretical maximum displacement that could be measured as the speed of light. If you stand outside on a clear night, you can see the Zodiac star constellations roll by. Marett determined that the laser beam should be aligned with the constellation Leo. As Leo rises in the east and sets in the west 12 hours later, the displacement results vary, with the maximum results early and late in the evening. Over 12 hours, the expected results were inconsistent. Over the next 1 ½ years, he modified the lab setup and finally discovered that as the laser gun heats up, the results change. Like Michelson, he deemed that he did not find what he was looking for.

In 2011 in a subparticle experiment, the CERN laboratory, which has the largest underground particle accelerator, clocked a neutrino at approximately 60 nanoseconds faster than the speed of light. 15,000 beams of neutrinos were fired approximately 500 miles to giant detectors. For the next 8 years they checked their results, could not find anything wrong with their method, and in 2019 published their results.

Figure 12: Rosette Nebula

Over the last half of the previous century telescopes, particularly, space telescopes, can view deep into the universe. As generational telescopes became more powerful, they viewed the light from what was thought to be the edge of the universe. The light had a red glow. This was caused by a Doppler shift of light frequencies from white to red as the Universe expands from its center; presumably the site of the big bang.

If an ambulance with a siren approaches you, sound waves get compressed into shorter wavelengths causing a higher pitched sound. As the ambulance zooms past your position, immediately there is a lower pitched sound as the waves traveling away from you get stretched. The same is true for light. Light traveling away from the Milky Way gets stretched and glows red. The amount of the shift determines its speed.

$$(23) \qquad Doppler\ Red\ Shift = \frac{Natural\ Light + Velocity}{Natural\ Light - Velocity}$$

By taking pictures of bright objects in space at different times, the red shift or the change in color can be measured over time and the speed of the object determined. Over the last half of the twentieth century the speed of distant objects increased as telescopes became more powerful. Each report was viewed with incredible skepticism. Galaxies were falling away at 90% the speed of light, then 92%, then 95%, and then an unbelievable 98% of the speed of light.

Nowadays, there are other ways to calculate the red shift. A cosmological red shift is a measure of how stationary light objects separate from one another as the Universe expands. This is like a hot-air balloon where the dots on the surface become more distant as the balloon expands. There now exists a scale measure that states if the Doppler Effect equals 1.4, the edge of the universe travels at the speed of light. In some cases in our century, galactical objects are traveling at the scale speed of 8, well exceeding the speed of light.

Some observers have stated that at the edges the speed of light slows down simply because of a long journey. A slower speed might mean less light and less of the red spectrum, but it would also mean less of a velocity difference among samples and, therefore, less measured speed. Another explanation has to do with the dark matter of the Universe. The current projection is that our universe is made up of 25% dark matter that cannot be seen by our telescopes. This dark matter is not organized and, therefore, has no measured gravity. Bright stars and nebula have gravity and tend to hold their position in the expanding Universe. As light travels back to our cameras from the beginning of time, the residue of the disappearing matter can be measured. While astronomers assumed that the expansion of the Universe was slowing down, measurements of dark matter phenomena show that Universe expansion is accelerating. Hence, these incredible speeds are forecasted from this acceleration given the current measured depth of the Universe. Compared with these measurements, celestial bodies with gravity are standing still.

One last hypothesis, the deep space temperature is 2 degrees Kelvin; that is, 2 degrees above absolute zero. Perhaps at the edges, the universe is colder. Light has fewer medium to travel through and therefore less friction or resistance. Perhaps in a differently configured world, light travels faster.

Chapter 4: The Exponential

Of all the irrational functions that mathematicians deal with, the exponential function (e) has got to be their favorite.

$$(24) \qquad \frac{d}{dt}e^t = e^t \ \ and \ \ \int e^t = e^t$$

Life does not get any better than this. The differential of any point along the curve is equal to the existing area under the curve. Potential energy is equal to kinetic energy. The geometric mean is equal to the exponential function. What could be cleaner? There is no residue. The Universe is a happy place.

The problem is that differentiation and integration work with continuous functions where the number of samples is infinite. For example, consider compounding:

$$(25) \qquad R = P\left(1 + \frac{r}{n}\right)^{nt} \qquad where$$

- R = return or result of compounding
- P = principle or initial investment
- r = rate of return or interest rate
- n = number of samples or number of time periods
- t = time unit

The problem to consider is when the number of samples (n) approaches infinity. Let us assume an initial investment (P) of 1 and a rate of return (r) of 100%.

$$(26) \qquad \mathrm{Lim}_{n \to \infty}\left(1 + \frac{1}{n}\right)^n = 2.71828\ldots = e$$

Unfortunately, no matter how vast, the number of objects in the Universe is finite. The exponential in all of its irrational digits is unattainable. Hopefully, there must be a discrete universal constant whereby difference equations and summations conserve energy. The ancients gave us many clues. From the Gospel of John, Chapter 1, Verse 1:

"In the Beginning was the Word, and the Word was with God, and the Word was God."

The Greek word "logos" is translated as "word", but there are a dozen other associations including the word "knowledge". Given in number theory that God equals unity, this lets us set up the following equation:

(27) $$W = \frac{1}{W} + 1 \qquad where$$

- W = the Word
- $\frac{1}{W}$ = the Word within God
- 1 = God

Solving:

(28) $$W^2 - W - 1 = 0 \quad W = Golden\ Ratio = \emptyset =$$
$$1.61759\ldots$$

This result is God's universal constant, which can also be expressed as its inverse 0.61759.

Figure 13: Da Vinci's Vitruvian Man

This 1487 drawing by Leonardo da Vinci of the Vitruvian Man of perfect proportions demonstrates how the Golden Ratio affects our everyday world. The ratio of the arm's length separated by the elbow is the Golden Ratio. The ratio of the belly button to the height of the man is the Golden Ratio. Your finger joints have Golden Ratio proportions. All trees have Golden Ratio proportions. The spiral of a sea shell has a Golden Ratio geometric mean.

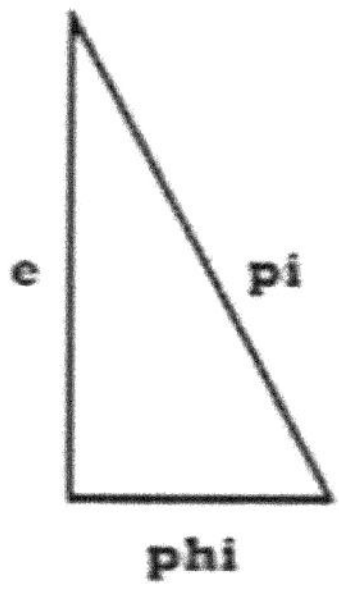

The orthogonal triangle shown relates the three most well known universal constants; well, almost. It is close. Using the Pythagorean Theorem, there is a 0.7% error with a hypotenuse value of 3.163397.

Figure 14: Pythagorian Universal Constants

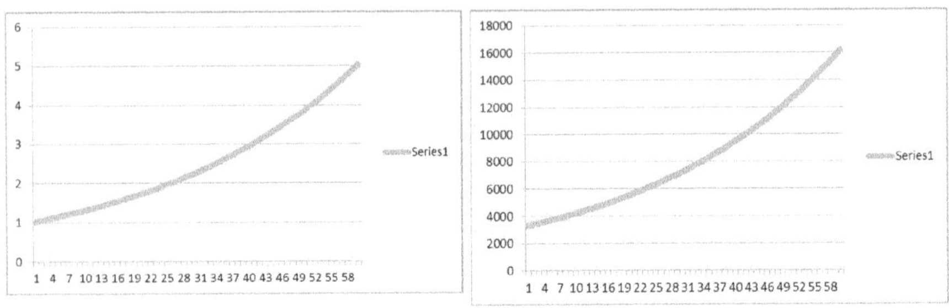

Figure 15: Exponential Results

The graphs above are compounding rate charts from simulations designed to explore the three universal constants: e, $\varnothing$, and π. The graphs show that compounding yields exponential results. For each of the three simulations, there are 360 samples or time periods (n). The principal (P) was 1 unit or dollar and the goal was to use the universal constants as the rate of return (r) for each simulation. The graph on the left shows the first 60 samples of e. The graph on the right showing the last 60 samples is identical to the graph on the left. The magnitudes in an investment sense are unimportant since this is a controlled experiment.

The results are tuned simulations. The three rates of return were tweaked to yield more favorable results. Let us look at the results first.

351	12698.808	272.096	46906.93
352	13045.315	276.476	48366.58
353	13401.278	280.928	49871.65
354	13766.953	285.451	51423.55
355	14142.607	290.046	53023.75
356	14528.511	294.716	54673.74
357	14924.945	299.461	56375.08
358	15332.196	304.283	58129.36
359	15750.560	309.182	59938.23
360	16180.339	314.159	61803.39
	Result1	Result2	Result3

$$\text{Result1} = 10000 * \varnothing$$
$$\text{Result2} = 100 * \pi$$
$$\text{Result3} = 10000 * 1/\varnothing$$

After taking into account the initial principal (P), the objective was to generate a result (R) using the three universal constants. The rate of return (r) values were tuned by a small percentage to yield these exact values. They were compounded 360 times (n). Yes, there are 360 degrees in a circle. Here are the three rates of return:

$$\text{Rate of Return1} = 2.728662 \cong e \quad \text{with a 0.38\% error}$$
$$\text{Rate of Return2} = 1.61002 \cong \varnothing \quad \text{with a 0.50\% error}$$
$$\text{Rate of Return3} = 3.11180 \cong \pi \quad \text{with a 0.96\% error}$$

Given the small error percentages, the question becomes: Are e, $\varnothing$, and π related? In 1996, Science magazine published an article about the "sinuosity" of rivers. The term sinuosity is a measure of how many bends a river has for a given distance. 496 rivers were in the study, and the average sinuosity was measured as 1.94.

$$(29) \qquad\qquad \frac{\pi}{\varnothing} = 1.94$$

Any fundamental frequency or pitch in nature is filled with overtones that have frequencies that are whole number proportions of the original scale. For example, if you hold the C-note down on a piano without striking it, and strike a lower C, then you will hear both notes and perhaps more faintly a few other notes.

Chapter 5: Singularity

Figure 17: Escher's Ascending and Descending

Perhaps it is too early in this discussion to call science a paradox, but here we are. The famous M. C. Escher drawing asks the question: Are you moving up the staircase or down? Which came first, the chicken or the egg? No matter how you move on the staircase, you wind up at the same place. You are part of an infinite loop. Your destination is a singularity – a location where your efforts become infinite.

In mathematics, an infinity (∞) usually is created or assumed when you divide by zero. Also, a function evaluated with infinite samples that can be described as divergent will climb or fall to infinity.

(30) $\lim_{n \to \infty} e^t = \infty$ *is divergent* $\lim_{n \to \infty} e^{-t} = 0$ *is convergent*

Many times, an infinity is the result of trying to rationalize a two dimensional problem in one dimension. You need to try a higher dimension, or at least think out of the box. The chicken and the egg have the same DNA. The staircase has only one direction (result).

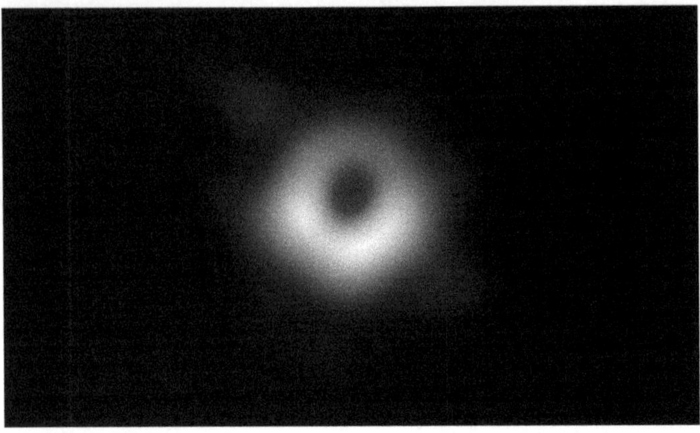

Figure 18: NASA First Black Hole Image (2019)

The creation of a black hole is based on a point singularity in the space-time continuum at the center of the hole. That is right; mathematically, it is a point and not a mass. The picture above was released in 2019 as the first photograph ever of the phenomena. A black hole consists of a black oval center and a bright ring around the hole called the event horizon. The gravity is so strong at the singularity due to the unseen condensed mass that it attracts matter and objects from nearby space. Light is bent toward the singularity creating the darkness. As the matter enters the hole, it is subject to strong forces that can rip chemicals and gases apart. Therefore, some antimatter particles enter the hole and destroy matter. Any collision between matter and antimatter emits photons of light. Eventually as matter is reduced, the black hole shrinks, gets hotter and light escapes the event horizon and looks like a laser flashlight in the sky.

A void or black hole might be created by a massive collision. Let us assume that the moon and the Earth are barreling toward each other. The closer they get, the stronger the gravitational force.

$$(31) \quad F = G_u \cdot \frac{m_1 m_2}{d^2} \quad where$$

- F = resultant force of two gravities
- G_u = universal gravitational constant = $6.674 \times 10^{-11}\ m^3 \cdot kg^{-1} \cdot s^{-2}$
- m_1 = mass of the earth = 5.9724×10^{24} kg
- m_2 = mass of the moon = 7.4×10^{22} kg
- d = distance between earth and moon

As the distance between the two bodies approaches zero, the resultant force approaches infinity. The equation indicates a force infinitely large; the only conclusion is that the Universe will disappear. The folly is that distance is measured from the center of each mass and not their surfaces.

$$(32) \quad \min(d) = r_1 + r_2 \quad where$$

- r_1 = earth radius = 6371 km
- r_2 = moon radius = 1737.5 km   therefore

$$(33) \quad F = 6.674 \times 10^{-11} \cdot \frac{5.724 \cdot 10^{24} \cdot 7.4 \cdot 10^{22}}{(6371 + 1737.5)^2} = 3.4864 x 10^{30}\ newtons$$

Over a distance of 1 meter, energy of $3.4864 x 10^{30}$ joules will create quite an explosion. However, the sun emits $3.86 x 10^{26}$ joules of energy every instance. The Universe is hardly in jeopardy.

Rearranging, the force (F) Equation (31) to assume that the radius of the Earth is much bigger than a second mass, in this case a spaceship, then:

$$(34) \quad F = \frac{G_u m_1 m_2}{r^2} \quad and \quad PE = \frac{G_u m_1 m_2}{r} \quad where$$

- PE = potential energy
- r = radius of the earth = 6371 km

To calculate the escape velocity of the spaceship from the Earth's gravity, by the conservation of energy, kinetic energy (KE) equals potential energy (PE) at the surface of the Earth.

$$(35) \qquad KE = \frac{1}{2} m_2 v^2 = PE = \frac{G_u m_1 m_2}{r}$$

$$(36) \qquad v = \sqrt{\frac{2 G_u m_1}{r}} = escape\ velocity$$

$$(37) \qquad v = \sqrt{2 * 6.674x10^{-11} * \frac{5.9724x10^{24}}{(6.371x10^3)}} = \sqrt{12.50x10^7}$$

$$(38) \qquad v = 11.18\frac{km}{sec} = 6.947\ miles/sec = 25,009\ miles/hr$$

That velocity will break all vehicle speed records on Earth but is no-where close to the speed of light. What if the Earth collapsed and became a black hole? Then no light must escape past the event horizon. In the velocity equation (36), we can substitute for velocity with the speed of light (c) and calculate the new radius of the Earth.

$$(39) \quad r_{new} = \frac{2 G_u m_1}{c^2} = 2 * 6.674x10^{-11} * \frac{5.9724x10^{24}}{(2.99702547x10^8)^2}$$

$$(40) \quad r_{new} = 8.875x10^{-3}\ m = 8.875\ mm$$

The Earth has collapsed to the size of a small marble and has disappeared as the center of a black hole.

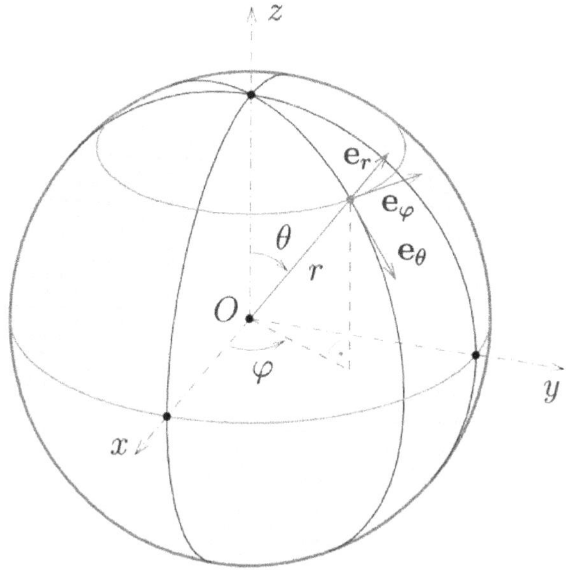

Figure 19: Space-Time Spherical Coordinates

With relativity, space-time distance is not measured with a straight line but is measured on the surface of a sphere in four dimensions: r,θ,$\emptyset$,and t.

(41) $\quad ds^2 = -\left(1 - \frac{2G_u m}{r}\right)dt^2 + \frac{dr^2}{\left(1 - \frac{2G_u m}{r}\right)} + r^2(d\theta^2 + \sin^2\theta d\emptyset^2)$ where

- $2G_u m = Schwarzschild\ radius\ (r_s)\ of\ a\ black\ hole$
- $ds = (space - time)\ distance\ between\ 2\ points\ on\ a\ sphere$
- $r = radius\ of\ a\ sphere$

Again, the distance on the sphere is faced with the anomaly of having the radius of the sphere disappear to zero causing an infinite distance. Or the distance between two points could be zero causing the black hole radius to be infinite. This does not happen because even an electron has a mass of $9.109 x 10^{-31} kg$. Even if a positron with an electron, the energy lost as heat would equal 0.511 joules. This is far less than the 10^{30}

joules of an Earth-moon collision, which in turn is far less than a supernova from a giant sun.

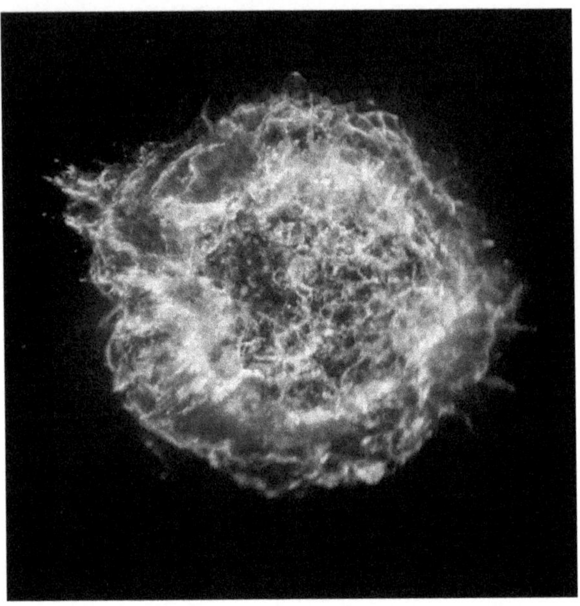

Figure 20: Cassiopeia - A Supernova Remnant

Stars 20 to 50 times the size of our sun form heavy metals at their core and eventually run out of hydrogen and helium. Without these gases, nuclear fusion stops pushing outward, and the gravity of the heavier elements takes over. These dying stars expire in the most violent explosions known in the Universe. In our Milky Way galaxy, a supernova happens every 50 years. If the mass of the star is 20 to 50 times the size of our sun, then the star eventually will collapse and form a black hole. At the instant of collapse, the star could have 100,000 times the mass of our star. Although still small, the collapse would not result in a singularity.

At some point, the science of black holes becomes conjecture. One of the most intriguing conjectures is that something is formed out of nothing. Using the law of energy conservation and starting with no energy, an electron (matter) and a positron (antimatter) are formed and then they come together to annihilate one another. Hence, the energy is still zero. If this spontaneous event happens on the event horizon of a black hole, then because of strong gravity forces, their positions get stretched. The positron enters the black hole and destroys a particle of matter. The electron stays outside the event horizon and adds to the matter of the Universe. Hence, the universe continuously expands.

Adding to the complexity of black holes is the Bekenstein equation for entropy (S) of a black hole. The Second Law of Thermodynamics states:

(42) $\qquad S = \frac{\Delta Q}{T} \qquad$ *where*

- $S =$
  entropy or the amount of potential energy lost in a dying system in joules per Kelvin
- $\Delta Q = $ *the transfer of heat loss in joules*
- $T =$
  the final or equilibrium temperature in degrees Kelvin

(43) $\qquad$ Bekenstein equation

$$\text{Max}(S) = \Delta S = \frac{4\pi k G_u m^2}{hc} \qquad \textit{where}$$

- $\Delta S = $ *the increase in entropy*
- $k = $ *Boltzmann's constant*
- $h = $ *Plank's constant*

Substituting the Schwarzschild radius of the black hole from Equation 41:

(44) $\qquad \Delta S = \frac{8\pi k r_s m}{hc}$

The change in area of a sphere is:

(45) $\Delta A = 4\pi r_s^2$ *therefore*

(46) $\Delta S = \dfrac{Akm}{r_s hc}$

One further substitution from quantum mechanics:

(47) $E = \dfrac{hc}{\lambda}$ *where λ = the wavelength of light*

Equating lambda (λ) with the Schwarzschild radius (r_s), which is the wavelength of light distance to the event horizon:

(48) $\Delta S = \dfrac{Akm}{E}$

When a black hole captures matter, the energy inside is increased. If a photon of light (λ) enters the black hole, the surface area of the black hole is increased and the maximum entropy is increased. Keep in mind that the increase in entropy leads to a potential energy maximum value for when the black hole dies. As the size of the black hole increases, real entropy is zero because there has been no decay. The Second Law of Thermodynamics describes increasing entropy as living systems die. As we shall see, this is not the case for a black hole.

Equation 42, the thermodynamics equation states:

(49) $\Delta E = T \, \Delta S$

(50) $\Delta E = \dfrac{hc}{\lambda}$ *for 1 photon*

Equating 1 photon as the basic unit for the change in entropy (ΔS) and substituting for r_s (Equation 41):

$$(51) \qquad T = \frac{hc}{\lambda} = \frac{hc}{r_s} = \frac{hc^3}{2G_u m}$$

This equation as applied to black holes was first thought of by Stephen Hawkings, the father of big bang radiation. Steven Hawkings was a student of Roger Penrose, the British mathematician. The equation describes the loss of photons or the shrinking of a black hole as an increase in temperature. It also describes the increasing temperature with the decrease of the black hole mass or size. Entropy is always associated with a decrease in temperature, down to absolute zero if necessary. Yet an increase in temperature suggests that the maximum entropy has increased, but thermodynamics says that entropy is inversely proportional to temperature. The equations hold but the logic does not. A dying system should cool and not be hot. What is "expounding" from the black hole is life, not death. It is as if the invisible universe has scientific laws that are opposite of the visible universe.

Figure 21: Rocket Launch

Lastly, a rocketship standing on the surface is not destined to fall through the Earth's center. However, when the Earth collapses to the size of a small marble as the center of a black hole, the gravitational forces are so large, on the order of $10^{11} kilometers/sec^2$, that covalent bonding, weak nuclear forces, strong nuclear forces, or any force on top of the marble like a spaceship or a penny would crush the marble. Hence, the center of the black hole becomes a singularity, a dimensionless point where the laws of physics no longer hold.

One argument against this view of black hole singularities becomes the following question: What happened to the heat friction caused by smashing quantum matter into the size of a marble? We are probably talking heat proportional to the energy of the collapsing giant star. The Plank's constant and quantum energy levels of electron orbits have been torn apart. Can any physicist be sure that the mass of a marble that equals the weight of the Earth would not fight back to keep its shape?

On October 6, 2020, Roger Penrose, the famous British mathematician, won the Nobel Prize in Physics for a paper he wrote in 1965 entitled, *Gravitational collapse and space-time singularities*. The conclusion was that there is a theoretical singularity inside a black hole. The mathematics stems from the tensor equations of general relativity, which are difficult to solve. However, there is a unique topological aspect to Penrose's theory.

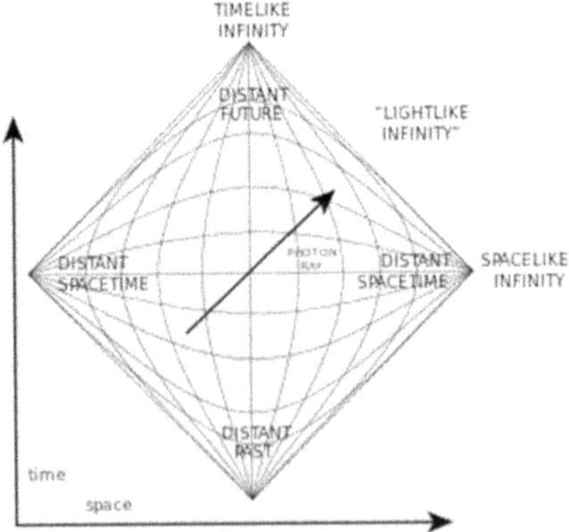

Figure 22: Singularities in Space-Time

Einstein's general theory of relativity models the gravitational coordinates of curved space in space-time coordinates. Roughly put, Penrose reasoned that gravity is so strong inside a black hole, all light from any direction captured by the gravity must intersect at a singularity, a central point, inside a black hole. The diagram points to four conclusions: There is no space, time no longer exists, there is no length to be measured, and the gravitational force is infinite. In other words, the general theory of relativity no longer applies.

The logical conclusions that Penrose reached produced a paradox. If Penrose is right then the general theory of relativity is wrong. Yet Penrose derived his results from the general relativity equations. This suggests that we must go up a dimension and think outside of the box in order to resolve the paradox.

The simplest argument is that time does not exist. Time is not a parameter. It is a unit of motion just like kilograms are a unit of weight. If there is no motion, there are no time units.

$$(52) \quad E = mc^2 = Einstein's\ equation\ for\ potential\ energy\ (PE)$$

$$(53)\ Alternatively,\ E = (PE_{max} - PE_{min})V \quad where$$

- $V = volume$

$$(54) \quad m = \left(\frac{(PE_{max} - PE_{min})V}{c^2}\right) \quad where \quad PE_{max} \gg PE_{min}$$

Simply put, if mass (m) is not infinite, gravity is not infinite. As the speed of light is a constant, time as a parameter plays no factor.

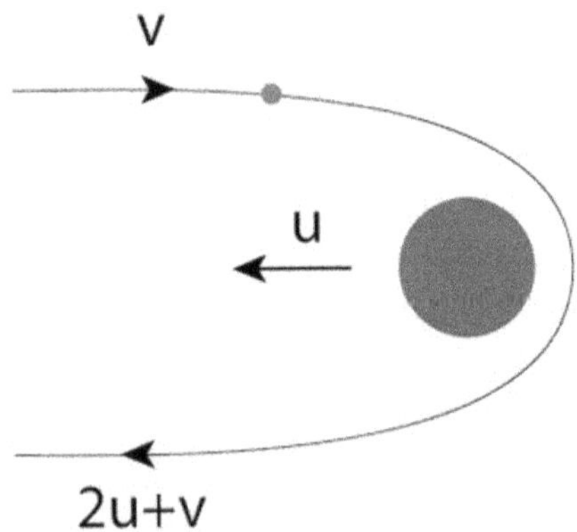

Figure 23: Gravity Slingshot Effect

Another perhaps more imaginary way to challenge Penrose's theory is the slingshot effect of gravity on a moving object. Give the correct direction of planetary rotation, the rotational speed of the planet can be added to the object's speed as it heads in a different direction. Photons travel at the speed of light. Given that gravity of a black hole is assumed to be infinity, either photons are traveling faster than the speed of light or the potential energy and therefore mass approaches infinity. This creates some highly unusual equations given that the velocity (v) of the photon beam is greater than the speed of light (c). Near the speed of light, Einstein's energy equation is rewritten as follows:

$$(55) \qquad E = \frac{mc^2}{\sqrt{1-\frac{v^2}{c^2}}}$$

$$(56) \qquad p = momentum = \frac{mv}{\sqrt{1-\frac{v^2}{c^2}}}$$

$$(57) \qquad \Delta t = change\ in\ time = \frac{\Delta t_0}{\sqrt{1-\frac{v^2}{c^2}}}$$

$$(58) \qquad L = distance = L_0 \sqrt{1-\frac{v^2}{c^2}}$$

If v > c, then all the equations above become negative and imaginary. There is no physicist that will accept these results. In addition, there is an indication that the photons are moving backwards in time.

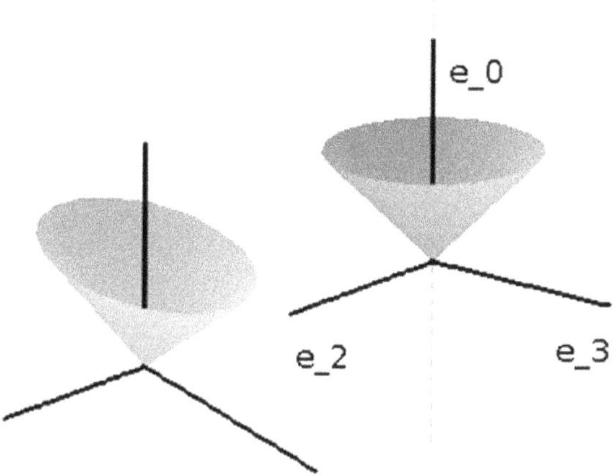

Figure 24: Godel Dust Cones

One of the most famous mathematicians of the 20th Century was Kurt Godel (1906-1978), a young friend and colleague of Einstein at Princeton University. In 1949, the Gödel metric introduced an exact solution of the Einstein field equations in which the identified stress–energy tensor contains two terms, the first representing the matter density of a homogeneous distribution of swirling dust particles, and the second associated with a nonzero cosmological constant. The cosmological constant is a creation of this theory. It is also known as the Gödel solution or Gödel universe.

Given that no matter can withstand the compression and gravity inside a black hole, at least some particles could be floating around as dust. The dust cones above might have enough gravity to capture photons into orbit around the cone. In any case, the cones are not a point. If the cones model reality, then there are no singularities as the cones represent mass and gravity is not infinite. In this solution, time does exist.

Chapter 6: Partial Order

Figure 25: Harold Hurst Treading a Nile River Outlet

In 50 years or so, the two most famous scientists in the 20th century might include a hydro-mechanical engineer treading down the Nile River. Engineers build things; they are looking for practical solutions. What this one discovered refutes many theories including Charles Darwin's Theory of Evolution.

Darwin's theory hypothesizes that evolution is a random process. The following table is a game that looks at randomness. You may recognize the digits in the first row with numbers as the digits of pi. In fact, the numbers listed are the first one hundred digits of pi, an irrational number. With the ten different numbers possible for a square, it is expected to take ten factorial combinations, a number equal to 3,628,800, to repeat a ten digit sequence. For this game, the numbers of pi are being used as a random number generator.

Other methods include using the serial numbers on 10 one dollar bills, if you want to make your own table.

a	b	c	d	e	f	g	h	i	j
3	1	4	1	5	9	2	6	5	3
5	8	9	7	9	3	2	3	8	4
6	2	6	4	3	3	8	3	2	7
9	5	10	2	8	8	4	1	9	7
1	6	9	3	9	9	3	7	5	1
10	5	8	2	10	9	7	4	9	4
4	5	9	2	3	10	7	8	1	6
4	10	6	2	8	6	2	10	8	9
9	8	6	2	8	10	3	4	8	2
5	3	4	2	1	1	7	10	6	7
A	B	C	D	E	F	G	H	I	J

Note: 0 digit set equal to 10

THE RANDOM NUMBER GAME

- From the top pick two columns.
- Starting with the number underneath your first column selection, advance to the right and down one row starting with the "a" column, equal to the number of squares specified by the number.
- With the number specified by your new square, advance to the right again.
- Repeat.
- When you have reached the bottom row, note the capital letter that you have landed on.
- For your selection of the second column, repeat the procedure.
- Compare the results from your two selected columns to the bottom row.

You should have landed on the same alphabetical square. In

fact, regardless of which column you start from, your counts will land you on the same bottom square. Why? You can create order out of randomness by .... using a CONSTANT PROCEDURE. In fact, you can change this procedure and create a new order, as long as you practice the same procedure. For a new procedure, the final bottom square is likely to change.

For example, try landing on a square, adding the number with the number of the square on the right, and then subtracting the number on the left from the sum. Occasionally, you will calculate a negative number and have to move backwards. Also, you may have to extend the alphabet of the last row.

Any number of procedures can be tried. If you are using multiplications or divisions, you probably will need a larger matrix of numbers. No matter how long your procedure takes or the path you start from, you will arrive at a single solution and create order out of seemingly noise, the randomness of something that used to be called manifest destiny. Individuals were not in control of their fate. The facts suggest otherwise.

The most famous game from Chaos Theory is that of the Sierpinski Gasket. For this game, you may need a computer to iterate a function 1,000 times or more.

THE SIERPINSKI TRIANGLE GAME

- Take three points in a plane to form a triangle, you need not draw it.
- Randomly select any point inside the triangle and consider that your current position.
- Randomly select any one of the three vertex points.
- Move half the distance from your current position to the selected vertex.
- Plot the current position (a mark).
- Repeat from third step.

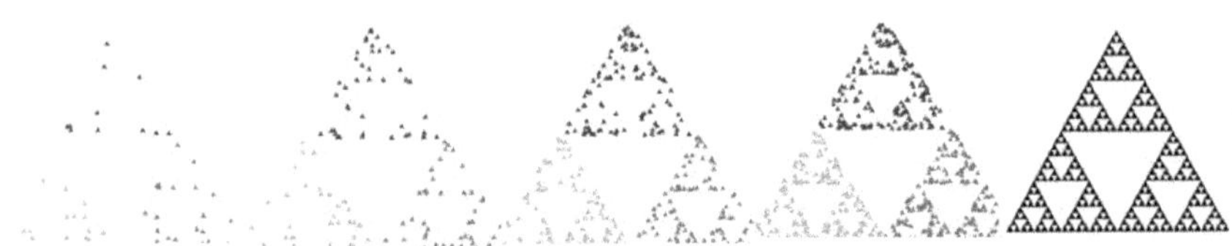

Figure 26: Sierpinski Triangle Panel

The more points that you mark between the initial triangle vertices, the clearer the image becomes. Waclaw Sierpinski, a Polish mathematician, invented the game in 1915. Today, it is called the chaos game because the smaller triangles formed are replicates of the original triangle. As is well known today, the world around us is fractal. If you peel the structure, eventually what you see is a replica of the structure. The triangle areas that are blank will be forever blank ... as long as you follow the same procedure.

Perhaps there might be skepticism as to how a game can be related to science let alone an artificial structure such as a Sierpinski Gasket. The picture below shows the bonding of electrons in a confined space on a metal surface. The process might be called algorithmic self-assembly. There is no metal stamp or computer controlled etching process. The electrons decide "randomly" where to go.

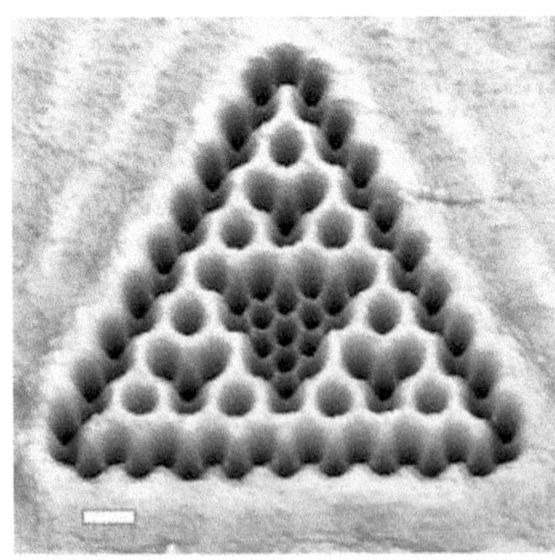

Figure 27: Electron Etching of Sierpinski Triangle

This particular study, *Design and characterization of electrons in a fractal geometry*, was done at Utrecht University and published in "Nature Science" on November 12, 2018.

"A relatively new topic in fractals is the quantum behaviour that emerges if you zoom in all the way to the scale of electrons. Using a quantum simulator, Utrecht physicists Sander Kempkes and Marlou Slot were able to build such a fractal out of electrons. The researchers made a 'muffin tin' in which the electrons would confine to a fractal shape, by placing carbon monoxide molecules in just the right shape on a copper background with a scanning tunneling microscope. The resulting triangular fractal shape in which the electrons were confined is called a Sierpiński triangle, which has a fractal dimension of 1.58. The researchers observed that the electrons in the triangle actually behave as if they live in 1.58 dimensions."

Benoit Mandelbroit , the father of Chaos Theory, invented the term "Fractal Dimension":

$$(59) \quad Fractal\ Dimension = \frac{\log(similar\ objects)}{\log(scalar\ dimensions)}$$

$$(60) \quad Fractal\ Dimension\ Sierpinski\ Gasket = \frac{\log(3)}{\log(2)} = 1.585$$

A Fractal Dimension essentially describes the complexity of an object when its algorithm for one level generates similar objects in a confined 2- or 3-dimensional space. Another algorithm for generating a Sierpinski Gasket is to draw an equilateral triangle, place a vertice halfway between each line, and connect the vertices:

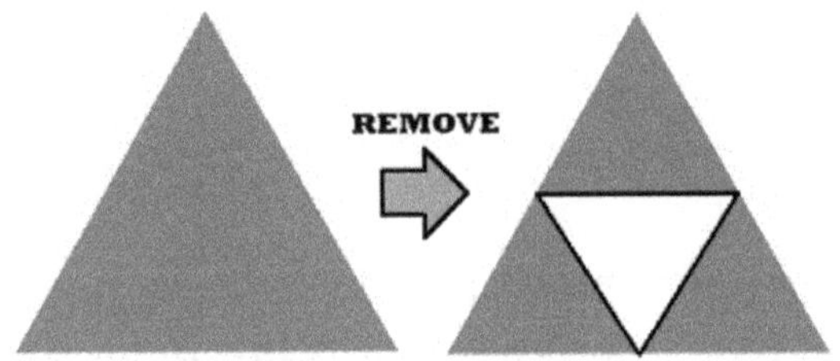

Figure 28: Sierpinski Triangle Method

After the removal of the inside triangle, 3 identical items in a 2-dimensional space remain. The Fractal Dimension will become increasingly important in future chapters. For now, the experiment at Utrecht University is only one of many experiments where the Sierpinski Gasket is bonded or etched on metal using a free-flowing chemical process.

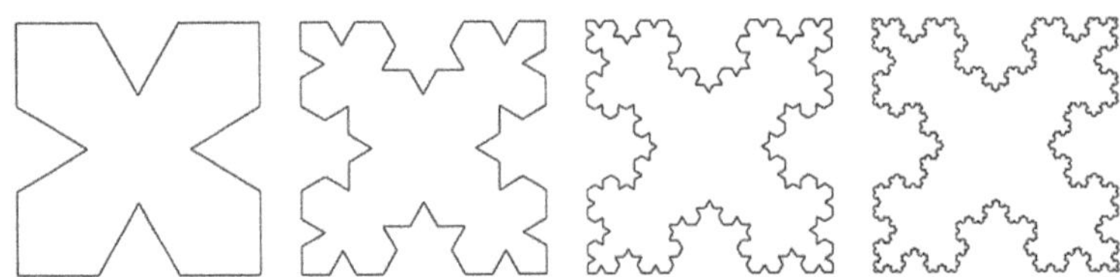

Figure 28: Koch Snowflake Panel

Another famous fractal structure is called the Koch Snowflake. Divide the sides into three parts and take out an equilateral triangle whose sides are equal to a third length of the sides. Continue the process until you have a snowflake. This simple structure has an advantage as a broadcast or TV antenna. The antenna on the next page is simply a level one (one iteration) Koch snowflake on only two of the four sides.

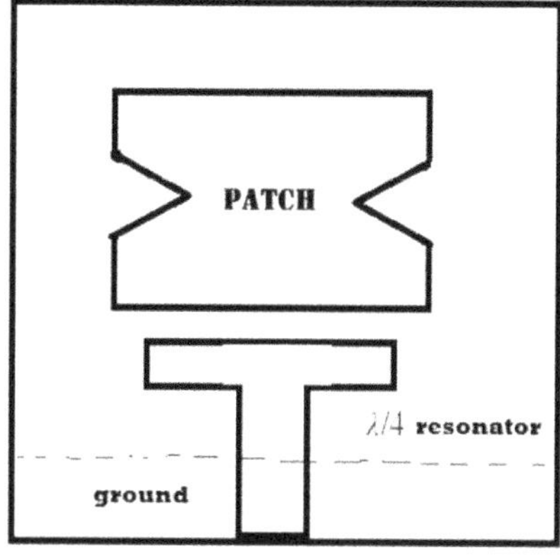

Figure 28: Koch Fractal Antenna

The advantage of a fractal antenna is that the fractal arms can be summed to produce a longer antenna in a shorter space. Sure enough for the four stubby arms of this antenna, the bandwidth is 35 times greater that the regular dipole rectangle. However, the gain is 67% less possibly because there is less material to resonate [Dr.Rekha Labade, KshitijaSanap, *Design and Analysis of Compact Fractal Patch antenna with Enhanced Bandwidth and Harmonic Suppression for WiMAX Applications*, International Journal for Research in Engineering Application & Management (IJREAM), 2018]. Not all fractal designs give similar results, but because of space requirements, the research continues.

For centuries, Egypt has depended on the Nile River to sustain its society. In the spring from the mountains in Iraq, the Nile River floods the basin spreading minerals and other nutrients across the land adding to the fertility of the soil. In more modern times with the increases in population, the amount of the Nile flow from the mountains can be the difference between flooding, abundance, and starvation for the populations of southern Egypt and surrounding countries such as Ethiopia. The solution is to build a dam for flooding and a water reservoir to supply water during times of drought.

Given the bar chart on the next page that shows the

amount of flow for every cubic kilometer of water over 137 years, the data possibly is random within the extremes of flow. Albert Einstein wrote of the Brownian Motion of electrons in 1926. This motion applies to any random process.

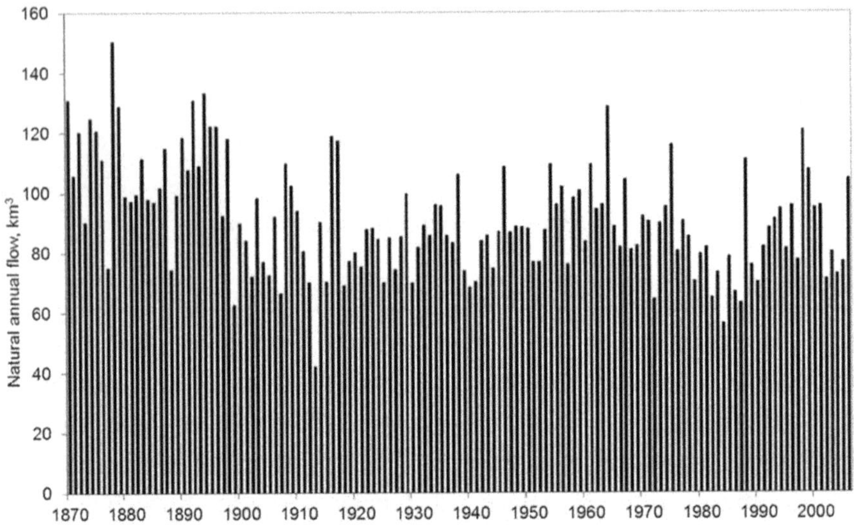

Figure 29: Nile River Water Flow Chart

(61) $$R = T^{0.5} \qquad where$$

- $R = distance\ (range)\ traveled$
- $T = time$
- $0.5 = index\ representing\ random\ walk$

Enter our hero who has been treading in the water. Harold Edwin Hurst worked on the water problems of the Nile River. He discovered that the water flow down the Nile was not random. There was a positive correlation to the flow that demonstrated that the water reservoir to support the populations did not need to be as big as randomness would suggest. His recommendations led to the Big Aswan Dam and water reservoir, which were completed in 1970.

Working with Einstein's equation, Hurst modified the equation to represent the self-similarity or persistence of the range of values in the study. Subsequently, this modification of randomness is now known as the Hurst Exponent.

$$(62) \qquad \left(\frac{R}{S}\right)_n = cn^H$$

$$(63) \qquad H = \frac{log\left(\left(\frac{R}{S}\right)_n\right)}{log(cn)} \qquad \text{where}$$

- $R = range(distance)$
- $S = standard\ deviation$
- $c = scaling\ constant$
- $H = Hurtz\ Exponent$
- $n = no\ of\ sample\ groups\ over\ a\ range\ of\ data$

The number of steps required to calculate the Hurst Exponent are significant. In essence, data are rescaled around zero by subtracting the mean from each point. Then the data are sampled or divided into groups of varying sizes and each individual group rescaled ratio of the range of the group (R) with respect to its standard deviation (S) is calculated. Given the number of sample groups (n) within the entire data for each standard size (for example; 2, 4, 8, 16, 32 or, 3, 9, 27, 81), the exponent equation (62) is simply the logarithm slope of the data. In other words, are the slopes of each group sampling similar to one another or are they dissimilar and therefore chaotic?

Given a self-similar alignment of slopes, there is a correlation or a consistency of slopes, the conclusion is that the data is persistent. For example, Hurst measured the persistency of the Nile River over 137 years with an average exponent value of 0.72. This means that following a good year of water flow, the water flow for the next year is 72 % likely to be good. This conclusion is an example of a dynamic partial order. The data are not random or dissimilar. Rather, the Nile River will always serve the population needs of the area with a little help in those lean years.

$H = 1$	The results are totally ordered
$H = 0.51$ to 0.99	The results are persistent; partially ordered
$H = 0.5$	The results are constant
$H = 0.01$ to 0.49	The results are anti-persistent; disordered
$H = 0$	The results are complete noise within the range

Harold Hurst was born in 1882 and published his initial technique around 1951. He studied other rivers such as the Great Lakes, Lake Albert, and Lake Victoria and in every case calculated an exponent of $H = 0.72$.

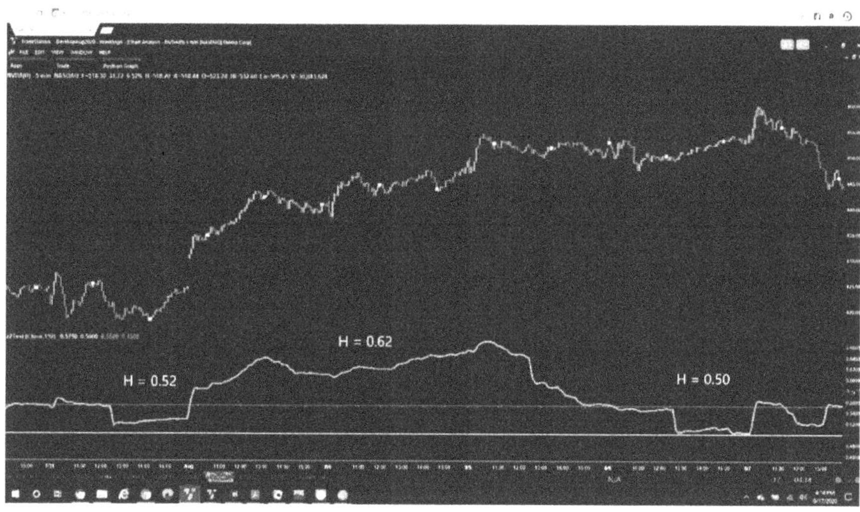

Figure 30: Changing Hurst Exponents for a Stock Chart

The stock chart above shows a Hurst Exponent routine that identifies two areas of random behavior and one large area of persistent behavior over a 5 day period. The random behavior features sideways motion, and the persistent behavior identifies a generally coherent gain over a period of three days. If the behavior was less than H < 0.50, an increasing anti-persistent behavior would look like noise; lots of short term ups and downs.

By identifying persistent behavior, Hurst presented an alternative to the notion that every process that science does not understand requires a random solution. In fact, Einstein's Brownian Motion becomes a special case. The likelihood that Darwin's evolution is random, that it can withstand the order evolving from any process solution to randomness is very small. Experts suggested that the Exxon Valdez oil spill in Alaskan waters was so large that nothing could live. Yet microbes evolved in the deep sea that could eat the oil. This is not random, but an ordered solution provided by nature to a human problem.

Chapter 7: The Harmony of the Spheres

Figure 31: The Planets

The moon may not be made of green cheese, but it not made of the minerals of Earth either. Either the moon or the Earth or both had to come from "outer space" before these celestial bodies began their elliptical orbits 93 million miles from the sun. Think about the possibility that the planet Earth traveling at 67,000 miles per hour, settled into an orbit around the sun 93 million miles away. If events are random, why not 75 million miles or maybe 125 million miles away?

Johannes Kepler (1571-1630) astronomer studied the orbits of the first six planets and developed three laws of planetary motion:

1. All planets move in elliptical orbits, with the sun at one focus.
2. A line joining any planet to the sun sweeps out equal areas in equal times.
3. The square of the period of any planet about the sun is proportional to the cube of the planet's mean distance from the sun.

The gravitational force (F_g) of the mass of the sun with respect to the mass of the planet is equivalent to the elliptical centripetal force (F_c) of the planet orbiting the sun. The Third Law of Kepler becomes:

(64) $F_g = F_c$

(65) $\dfrac{GMm}{r^2} = mw^2 r$ *where*

- $G = gravitational\ force$
- $M = mass\ of\ sun$
- $m = mass\ of\ planet$
- $r = radius\ of\ elliptical\ orbit\ (distance\ from\ the\ sun)$
- $w = angular\ frequency = \dfrac{2\pi}{T}$ *where* $T = time$

(1) $GM = \left(\dfrac{2\pi}{T}\right)^2 r^3$ *therefore* $GM = \dfrac{4\pi^2 r^3}{T^2}$

(2) $T^2 = 4\dfrac{4\pi^2 r^3}{GM}$ *therefore* $T^2 \propto r^3$

This famous proportion sometimes called the Law of Harmonies sets up the following table:

Planet	Period T (yr)	Average R (yr2/au3)	T2/R3 Distance (au)
Mercury	0.241	0.39	0.98
Venus	0.615	0.72	1.01
Earth	1.00	1.00	1.00
Mars	1.88	1.52	1.01
Jupiter	11.8	5.20	0.99
Saturn	29.5	9.54	1.00
Uranus	84.0	19.18	1.00
Neptune	165	30.06	1.00
Pluto	248	39.44	1.00

The almost perfect order for all planets established by the proportion means that the distance from the sun is no accident; it is not random. Possibly the only reason that all orbits are not perfect is that for a planet the orbit is not settled because the solar system is relatively young compared to the forecasted age of the Universe; 4.6 billion years compared to 13.8 billion years. In any event, these are revealing results as to systemic order.

Certainly music affects everyone, but why a good piece of music makes the heartbeat race is a mystery. So consider the following question: Can you sonically hear your family members over a cell phone? Unfortunately, this is a trick question. Of course, you may talk with family members on the phone every day successfully, but that is not the real question. Does the sound of a family member over the phone reach your eardrum? Can you hear them sonically?

Sound is frequency measured in hertz units. In general, the audible frequency range is considered to be from 20 hertz to 20,000 hertz. The sound from a phone emanates from a speaker and as cell phones are manufactured smaller, the speaker or speakers must be made smaller. Almost all cell phone speakers have a solid frequency response from 200 hertz to 5,000 hertz. Fortunately, female voices have a fundamental frequency centered around 250 hertz, so your aunt can be heard clearly. However, most men voices are centered between 100 to 125 hertz. Well below 200 hertz is not audible on a cell phone speaker. That means that the fundamental frequency of man's voice is not audible on your eardrum, yet you may talk clearly with your uncle every day. This is a mystery.

What distinguishes the sound of a man's voice say from the sound of a clarinet are called overtones. More specifically, the timbre or strength of each overtone distinguishes the sound. A sound is made from a fundamental frequency and a series of overtone frequencies that are higher pitched, have more frequency vibrations than the fundamental frequency. A man saying hello at a fundamental frequency of 100 hertz will also generate frequencies at 200 hertz, 300 hertz, 400 hertz, 533 hertz and so forth. These frequencies sonically are heard on the eardrum. Then something amazing happens. Your brain is familiar with the overtone series and automatically decodes the overtone frequencies and extrapolates the fundamental frequency of 100 hertz in your mind. Despite the fact that your eardrum did not receive the fundamental frequency sonically, you uncle clearly can be heard saying hello over the phone.

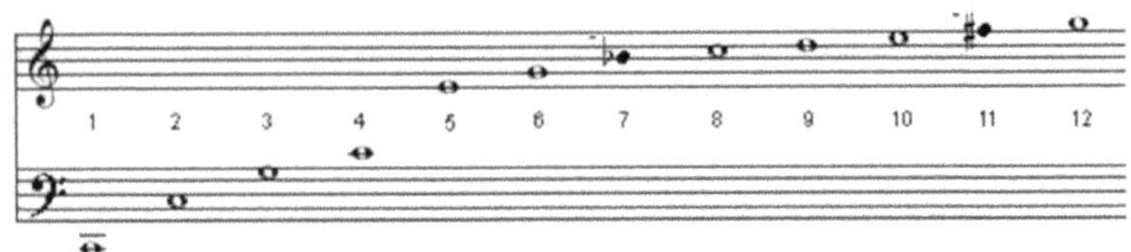

Figure 32: The Overtone Series Revisited

The lowest tone on the above music staff is a C note, on a piano the note is called C1 for the first scale octave comprising eight notes. The next note is a perfect octave above called C2. Ironically C2 is twice the frequency of C1. If you look at the numbering that 2 follows 1, then you can correlate the frequencies by multiplying the lowest tone by a ratio of 2 to 1. Continuing the numbering, multiply the C2 note by the ratio of 3 to 2 generates G2. Continuing the algorithm, the following notes are generated: E3, G3, weak note, C4, D4, E4, week note, G4. So far for the C scale, the white notes on the keyboard, the following notes have been generated: C, D, E, G, and back to C.

Interval	Ratio to Fundamental Just Scale
Unison	1.0000
Minor Second	25/24 = 1.0417
Major Second	9/8 = 1.1250
Minor Third	6/5 = 1.2000
Major Third	5/4 = 1.2500
Fourth	4/3 = 1.3333
Diminished Fifth	45/32 = 1.4063
Fifth	3/2 = 1.5000
Minor Sixth	8/5 = 1.6000
Major Sixth	5/3 = 1.6667
Minor Seventh	9/5 = 1.8000
Major Seventh	15/8 = 1.8750
Octave	2.0000

By continuing the overtone series beyond the first 12 notes, the missing scale notes of F, A, and B will be generated. This is Nature! Musical note frequencies are not arbitrary. They are fixed as overtones once the fundamental note, called the tonic note of a scale notation, is played.

The physical nature of the notes of a scale is that they are whole number ratios of the fundamental or tonic frequency. Musicians call Nature's tuning "just tuning". So let us generate the frequencies for the C scale starting with middle C on the piano. Middle C in our modern world has a frequency of approximately 262 hertz.

NOTE		C	D	E	F	G	A	B	C
INTERVAL		Unison	Maj 2nd	Maj 3rd	4th	5th	Maj 6th	Maj 7th	Octave
RATIO		1	9/8	5/4	4/3	3/2	5/3	15/8	2
FREQ.		262	294.75	327.5	349.33	393	436.67	491.25	524

Today, "Nature's tuning" is not quite so simple because musicians modulate between keys. Therefore, small frequency adjustments to notes had to be made in an equal way so that the note G, for example, has the same frequency in the C scale, as it has in the F scale or any scale for which it is a member. This modern tuning is called "well-tempered" tuning.

While not conclusive, the suggestion is that sound frequencies, indeed all frequencies, have an effect on our bodies and our minds that we may not be able to control or understand. Perhaps the most significant music in our world is "The Music of the Spheres". The Earth is 93 million miles from the sun. By having the scientific community assign this distance as 1 Astrological Unit (A.U.), the relative distances of the other planets from the sun can be calculated as in the table on the next page.

Body	Actual mean distance in A.U.	Approximate Whole-Number Ratio	Decimal Ratio
Mercury	0.39	1	1
Venus	0.72	15/8	1.846
Earth	1.00	4/3	1.389
Mars	1.52	3/2	1.52
Asteroid Belt	2.77	15/8	1.824
Jupiter	5.20	15/8	1.877
Saturn	9.54	15/8	1.835
Uranus	19.19	2	2.012
Neptune	30.05	3/2	1.566
Pluto	39.54	4/3	1.316

To the right of each distance, an approximate whole number ratio is calculated based on dividing a spheres distance by the previous sphere's distance from the sun. Note that these distances are for orbits of a relatively young solar system and may still have some adjustments to make. The results are pretty spectacular as these ratios all belong to the ratios of our major scale – "The Music of the Spheres".

Pythagoras speculated that each of the planets, through their orbits, must produce a particular note according to its distance from an immovable center (Earth). At the time, the Earth was the center of the Universe. Just as differing the length of a string adjusts its pitch when the string vibrates, so these varying distances must produce different tones: the "Music of the Spheres", no less. At the time of Pythagoras, only six planets were known.

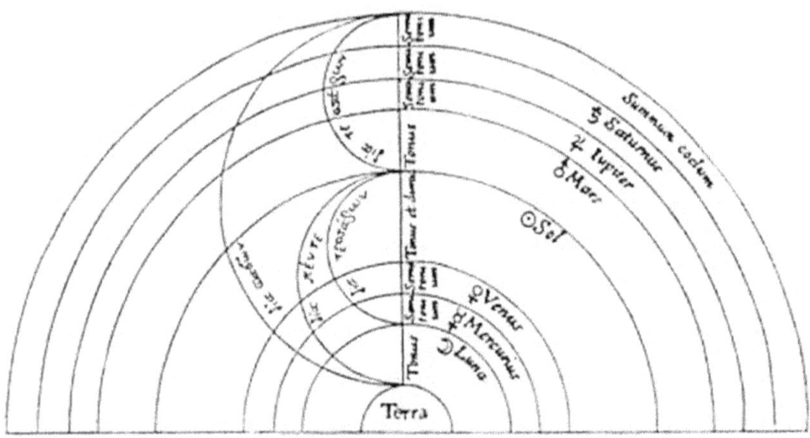

Figure 33: Pythagoras Music of the Spheres

The major tenant of the Pythagorean "Music of the Spheres" was the distance each planet travelled or orbited around the Earth as a perfect circle. This included the Moon as the first object, the six planets, the sun, and the fixed stars as the last object. As the reasoning was that celestial bodies that were farthest away travelled slower, then the musical pitch would be lower as distance increases.

Johannes Kepler used his Third Law to actually calculate the wavelength or period (frequency = f) of a planetary sound. From Equation 65:

$$(66) \qquad f = \frac{1}{T} = \frac{1}{4\pi} * \sqrt{GM/r^3}$$

In addition and for the first time, Kepler's Second Law clearly indicates that planetary orbits are elliptical. Therefore, the extreme distances from the sun indicate a range of musical notes for each planetary body. His Harmonicis Mundi treatise of 1619 that revealed his Third Law of planetary motion graphically shows his results.

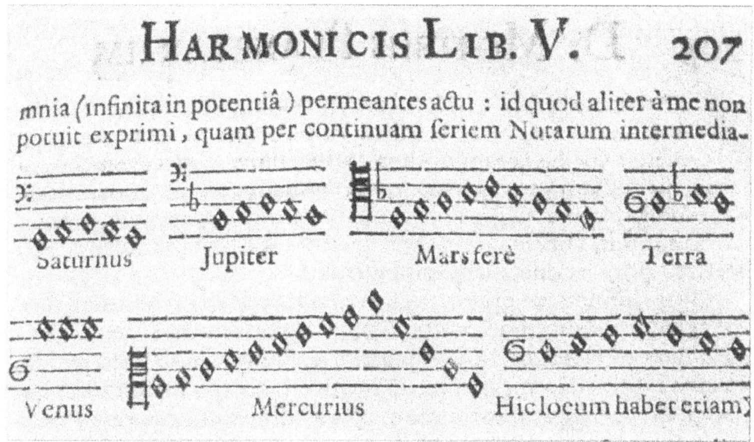

Figure 34: Kepler Harmonicus Mundi Music (1619)

Note that the orbit of Venus is close to a perfect circle; hence the repetition of the same note.

Music ratios then are serious business and not just for musicians. Our world is built out of these ratios, and there is one more number that is terribly significant for our world. By adding all the ratios and dividing by the number of celestial bodies, you find the average ratio distance between the spheres. That number is 1.617. Since the ratios are approximate, let us use the last column of our table with the more exact decimal ratios and try the same calculation. The more accurate calculations average to 1.618. As per Chapter 3, this is the Golden Ratio.

Figure 35: Milky Way Galaxy

The Earth's sun is one of 400 billion stars in the Milky Way Galaxy. It is about 26,000 light years from the center of the galaxy, which is about one-third of its radius. There is a black hole at the center that would fit inside the orbit of Mercury. The Milky Way has spiral arms that can be modelled by a Fibonacci sequence. This most famous of the sequences is constructed by adding the previous two numbers together and generating the next data point:

(67) *Fibonacci Sequence* $(1\ 1\ 2\ 3\ 5\ 8\ 13\ 21\ 34\ 55\ 89\ 144\ 233\ 377\ ...)$

(68) *Fibonacci Ratio* $= \dfrac{a_n}{a_{n-1}} = \dfrac{55}{34} = \dfrac{89}{55} = \dfrac{144}{89} = \dfrac{233}{144} = 1.618 =$

Golden Ratio

The fact that the increasing bend of a Milky Way spiral of stars can be modelled by a Fibonacci sequence is no random accident of Nature. The Milky Way is rotating and it takes approximately 220 million years for its outer spiral to make one rotation about the center. This is the local order of our galaxy.

Chapter 8: Solar Wind

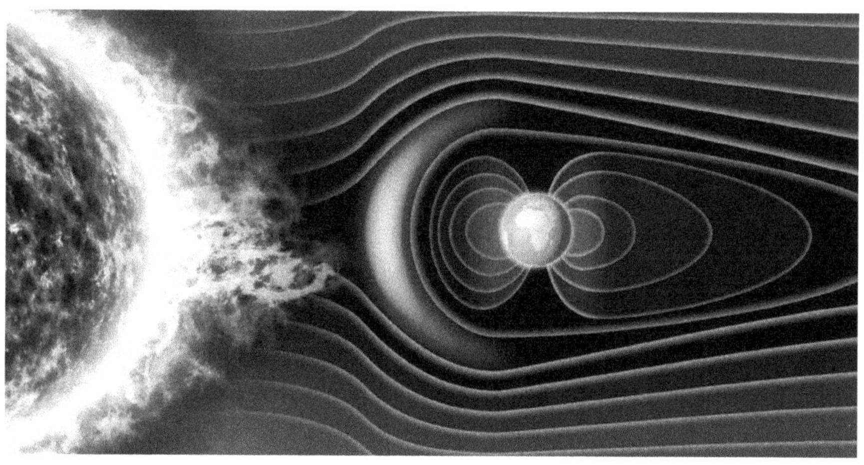

Figure 36: Solar Winds

In 1968, President Richard Nixon ordered the mining of Hiaphong Harbor off the coast of North Vietnam to cut off their naval supplies. He had hoped that this maneuver would bring the North Vietnamese to the table and end the war. While a naval ship watched the harbor, all of the submersible mines exploded for no apparent reason. The military explanation was that a large solar flare had penetrated the Earth's magnetic field and sent a current through the mines thus triggering the senseless explosions.

Figure 37: Northern Lights

A solar wind consists of charged particles of electrons and protons that have been discharged from the suns atmosphere by a solar flare. The gaseous exterior of the sun has a temperature of 1 million degrees Kelvin, and is easily able to split atoms apart. As they travel through space, the charged particles create an electric current and the current creates a magnetic field. When the solar wind bumps into the magnetic field of Earth, the collisions of particles emit photons creating the Northern Lights or the aurora borealis. These lights can appear in other regions of the Earth but the dipole effect created by a hot and liquid rotating core means that the lights are most likely to be seen at the poles. And the light show will be different every night depending on the interaction. Other planets such as Jupiter also experience the aurora borealis lights.

In general, the relationship between current and magnetism is explained by Maxwell's Equations. James Clerk Maxwell (1831 – 1879) was an English scientist and physicist. His Third Equation explains the dance between electricity and magnetism.

$$(69) \quad \nabla \times E = curl\, E = -\frac{\partial B}{\partial t} \quad\quad where$$

- $\nabla = curl = electric\, flux =$ *dimenional partial derivatives* $(\frac{\partial}{\partial x}, \frac{\partial}{\partial y}, \frac{\partial}{\partial z})$ *applied to the electric field*
- $E = electric\, field\, full\, of\, charges$
- $B = magnetic\, field$
- $\frac{\partial B}{\partial t} = change\, in\, magnetism\, flux\, 90\, degrees\, from\, the$ *electric field dimensions*
- $\nabla \times E = cross\, product\, of\, vectors$

An electric or magnetic field can be a circle or a rectangle or a football field. The players on the field are either electric charges or magnetic dipoles. The Third Maxwell Equation in differential form is the fundamental relationship between electricity and magnetism. A change in the electric field causes a change in the magnetic field and visa-versa. A change in the magnetic field causes a change in the electric field.

Ironically, your right hand and your thumb explain the equation better than a mathematical explanation. If there is an electric current of charged particles passing through a line or chord and your thumb points in the direction of the current, then the curl of your fingers illustrates the magnetic field generated that rotates around the electric line. Again, the roles can be reversed. If the line is magnetic, then a current is generated that curls around the line. These electromagnetic principles are at the heart of many devices such as transformers, power supplies, and magnetic imaging resonance machines for hospitals.

CURL RIGHT HAND RULE

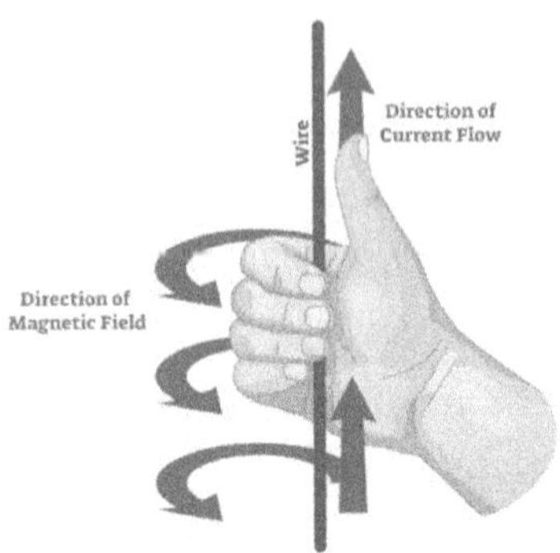

Figure 38: Curl Right Hand Rule

Resonance is a property that allows for a maximum efficiency or response due to the alignment of frequencies or whole number ratios. If you are between frequencies on an FM radio, what you hear is pink (audible) noise. When you tune the radio to a station frequency so that you hear the program, you are in syncopation with the resonant frequency of the radio station.

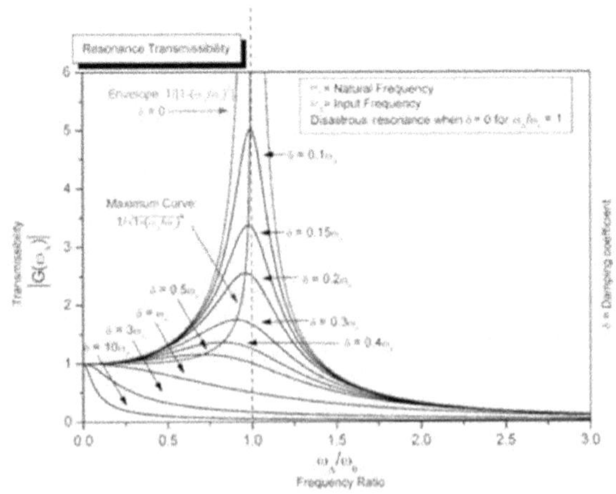

Figure 39: Resonance

Jupiter has many moons. The three largest moons – Io, Europa, and Ganymede – are approximately the size of our moon. Their orbits around Jupiter are spaced in perfect octave harmony: 1 to 1, 2 to 1, and 4 to 1. Once every eight orbits of Io, the three moons are aligned with Jupiter. Their strong gravitational field are in resonance with one another and affect the positions of the other moons.

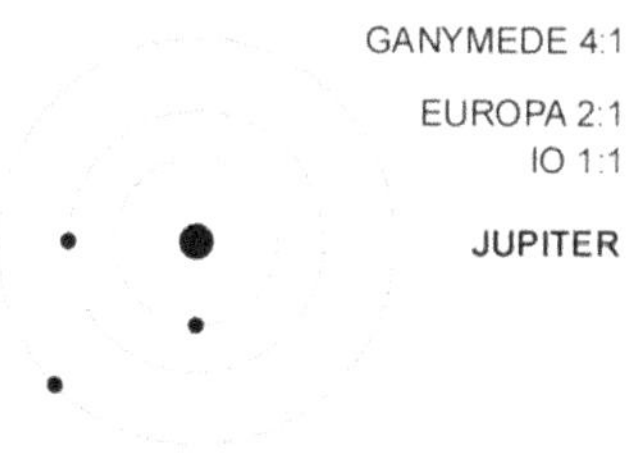

Figure 40: Moon Orbits around Jupiter

Jupiter has the strongest gravity of any planet because of its large mass. The supposition presumes that the Earth is protected from a meteor shower because of the gravity of Jupiter. Indeed, the feeling is that any solar system that supports life must have a larger planet to protect life. The Asteroid Belt between Mars and Jupiter is affected by Jupiter's gravity.

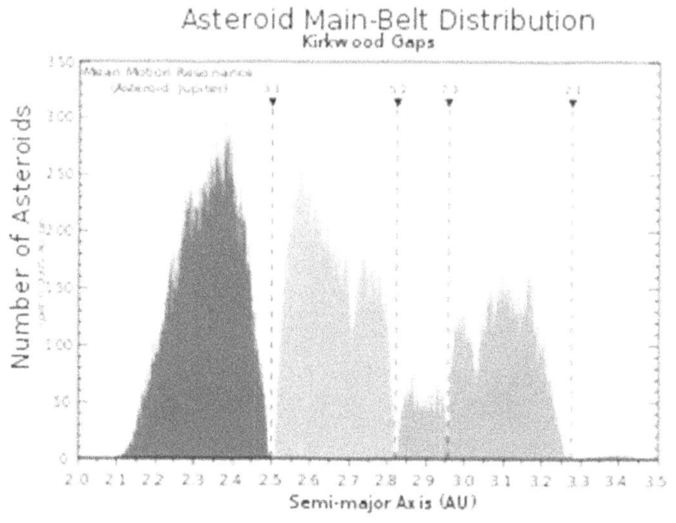

Figure 41: Densities of Asteroid Belt

The Asteroid Belt contains more than 100,000 meteors or asteroids. However, the previous graph shows that at certain whole-number ratios from the sun in astrological units, the number of asteroids is almost non-existent. These vacant spots that lie within the Asteroid Belt are due to the gravitational resonances emanating from Jupiter. The asteroids at these resonances have been pulled from the belt and either burn up in Jupiter's atmosphere, become part of the cloud of objects that follow Jupiter's orbit, or become another Jupiter moon. The present count stands at over 80 moons for Jupiter.

In 2009, the Kepler Space Satellite was launched. Its telescope observed a fixed field of 160,000 stars for ten years. The objective was to discover planets at least the size of Earth. By observing the star light being obstructed by a celestial body, the Kepler Satellite discovered approximately 2,700 planets. The satellite also observed the periodic rotation of the planets by their orbits.

Figure 42: Magellan Galaxies

The gaseous clouds formed in deep space presenting a light show are caused by stellar winds. In essence, stellar winds are solar winds emanating from suns such as red dwarfs that are much larger than our sun. Generally, these gaseous formations are hotbeds of new life; new solar systems.

In 1519, the explorer Ferdinand Magellan and his crew noticed a hazy blue patch in the southern sky. The blue and red formations are satellite galaxies of the Milky Way, and rotate around the Milky Way. The pink-tinged cloud, NGC 2014, is a glowing cloud of mostly hydrogen gas. It contains a cluster of hot young stars. The energetic radiation from these new stars strips electrons from the atoms within the surrounding hydrogen gas, ionizing them and producing a characteristic red glow. These formations are nicknamed after Magellan. The red ribbon is known as the Magellanic Stream and the blue cloud is known as Magellanic Cloud.

Massive young stars also produce powerful stellar winds that eventually cause the gas around them to disperse and stream away. The blue donut cluster is an example of this. A single brilliant and very hot star seems to have started this process, creating a cavity that appears encircled by a bubble-like structure called NGC 2020. The distinctive blueish color of this rather mysterious object is again created by radiation from the hot star - this time by ionizing oxygen instead of hydrogen.

The red cloud is only about 163,000 light-years from our galaxy and so is very close on a cosmic scale. Although very large, the red cloud contains less than one tenth of the mass of the Milky Way, and spans just 14,000 light-years. By contrast, the Milky Way covers some 100,000 light-years. Astronomers refer to NGC 2024 as an irregular dwarf galaxy; its irregularity, combined with its prominent central bar of stars, suggests that interactions with the Milky Way and the nearby blue galaxy NGC 2020 could have caused its chaotic shape.

Chapter 9: Olbers' Paradox

There are one hundred billion trillion stars in the observable Universe; a staggering number. If one were to stare at the Andromeda Galaxy, presumably you would observe one trillion stars. As viewed from Earth, the apparent size of the Andromeda Galaxy is 3.167 degrees by 1 degree, and 110,000 light years. The Andromeda Galaxy might be hiding two galaxies from our site within the spreading angle by distance, and the two galaxies might be hiding four galaxies and so on. Since photons should not loose magnitude in the vacuum of deep space, and since the total sum of light 200,000 light years away for two galaxies might be equivalent to the sum of light of one galaxy 100,000 light years away, why is outer space so dark? This is Olbers' Paradox.

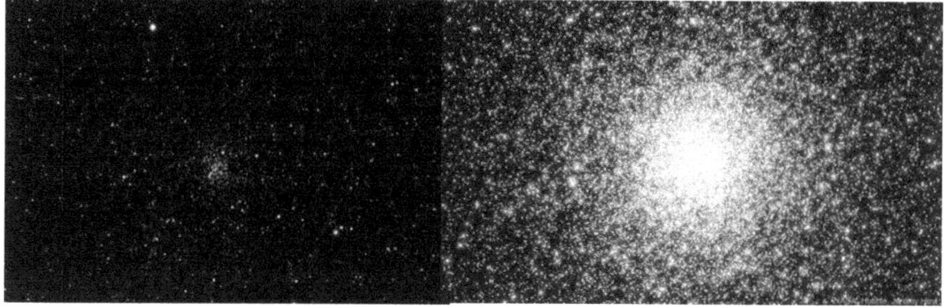

Figure 43: Stars in the Night Sky

Heinrich Olbers (1758 – 1840) was a German physician and a part-time astronomer. He discovered the asteroids Pallas and Vesta. His paradox about light was based on the philosophical observations of Isaac Newton. Newton believed in a static Universe; that is, a Universe with infinite stars and an infinite age. Every star was firmly fixed in the night sky. If these assumptions were true, then the night sky should be ablaze with light.

To help us with the paradox, let us consider the creation theories of the Universe in the order in which they were expounded:

1. Big-Bang Theory
2. Steady-State Theory
3. Fractal Generation
4. Pulse Theory

1. The Big Bang Theory

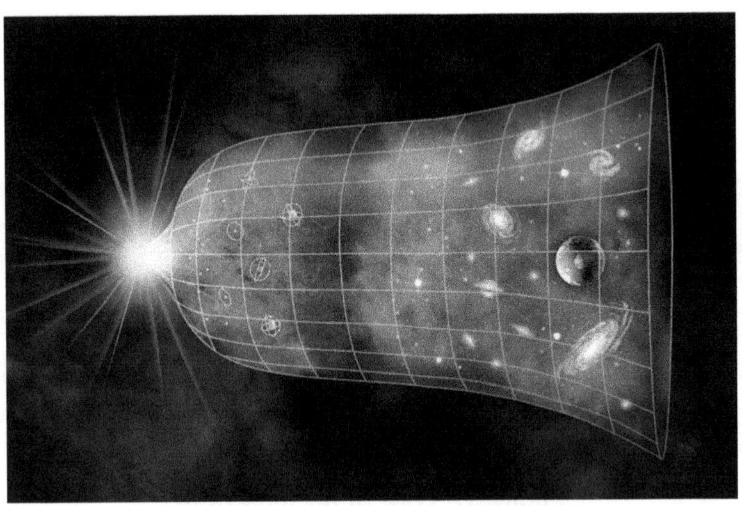

Figure 44: Big Bang through Time

The Big Bang Theory is an explosion at the beginning of time. All matter was densely packed in what is described mathematically as a singularity or a point. The mathematics is derived from Einstein's General Theory of Relativity. An antimatter particle got loose, attacked matter, and at a trillion degrees, caused an explosion flinging all particle ions into outer space. Around 400,000 years later, the Universe was cool enough to form atoms of hydrogen, some helium, and enough material to form the first hydrogen gas sun. At about one million years, the Universe was dense enough to form more complex stars and to begin to form galaxies. The Big Bang Theory is very popular with today's scientists because their telescopic red shift measurements show that the Universe is expanding; moving away from the center of the Universe.

There is a religious irony that in 1931, a Belgian Catholic priest Georges Lemaitre proposed the Big Bang Theory. He based his rationale on the red shift measurements of the Hubble Telescope that confirmed his theory of an expanding Universe. The theory was further strengthened when the use of radio telescopes discovered a uniformly dense radiation belt in deep space near the beginning of time. Radiation happens when charged particles that have no mass are detected. In the early Universe, there were no atoms because of the intense heat; only charged particles that would come together and form atoms as the Universe cooled.

Figure 45 Wormhole   *Figure 46: Quasar*

Like all science, the hypotheses are confirmed or denied based on current observations. All hypotheses run into trouble, and like a leaking boat, have to keep plugging the holes with interesting corollaries. The Big Bang Theory of everything has many complex parts. While not commonly discussed, Einstein's relativity equations for curved space lead to a singularity. Star Trek fans understand that traveling through a wormhole created by a black hole can be a short tunnel to somewhere else in space as space-time folds on itself. At the instant of creation, gravity is infinite meaning that space-time curvature reduces to a point. While this fact may be hard to believe, intuitively, simple arithmetic is not capable of a singularity. This will be explained more fully in Chapter 14 Ancient Geometry. In addition because of the infinite number of space-time folds, the center of the Universe, where presumably the Big Bang happened, cannot be predicted. Since an explosion radiates in all directions, the Big Bang Theory cannot explain the flatness of the Universe. The Universe rotates just like our Milky Way galaxy as essentially a flat disk.

Another interesting situation is the appearance of quasars in early time. Since light in the Universe takes millions of light years to reach Earth's telescopes, the early history of the Big Bang can be observed. A quasar is an extremely luminous light or ray of radiation that escapes the strong gravitational forces of a black hole. This only happens as the black hole is dying. As the black hole shrinks in size, it becomes hotter and the radiation is allowed to escape. Quasars only exist in early space-time. Their current red shift measurements state that quasars are traveling away from us at 80% the speed of light. They do not materialize in the time of our solar system. Therefore, how is it possible for black holes to exist in early space? It would take billions of years for a large sun, 100 times bigger than our solar system sun to form. In addition, a large sun at its core would form more complex atoms like iron that were simply not possible for the early hydrogen and helium stars.

Other facts seemingly have no explanation. By their own admission, scientists hypothesize that the laws of science do not exist in the fraction of a second of the explosion. In addition, the uniformness of early radiation does not seem to coincide with the expanding Universe. There should be packets of thinness in the radiation observed. Could it be that after billions of light-years, the measured red shift at the observable edge of the Universe is really light moving faster due to lowering temperatures?

2. Steady-State Theory

It was Fred Hoyle, British Astronomer, who coined the phrase the "Big Bang". Ironically, he and his colleagues Thomas Gold and Hermann Bondi formally proposed the Steady-State Theory in 1948. From his publications, Fred Hoyle's quotations explain his point of view:

"Life cannot have had a random beginning. ... The trouble is that there are about two thousand enzymes, and the chance of obtaining them all in a random trial is only one part in 10 to the 40,000 power, an outrageously small probability that could not be faced even if the whole Universe consisted of organic soup."

"Space isn't remote at all. It's only an hour's drive away if your car could go straight upwards."

"One [idea] was that the Universe started its life a finite time ago in a single huge explosion, and that the present expansion is a relic of the violence of this explosion. This big bang idea seemed to me to be unsatisfactory even before detailed examination showed that it leads to serious difficulties."

"Science is prediction, not explanation."

"There is a coherent plan to the Universe, though I don't know what it's a plan for."

"Things are the way they are because they were the way they were."

"Some super-calculating intellect must have designed the properties of the carbon atom, otherwise the chance of my finding such an atom through the blind forces of nature would be utterly minuscule."

"The big bang theory requires a recent origin of the Universe that openly invites the concept of creation."

"Perhaps the most majestic feature of our whole existence is that while our intelligences are powerful enough to penetrate deeply into the evolution of this quite incredible Universe, we still have not the smallest clue to our own fate."

"It seems to be a characteristic of all great work that its creators wear a cloak of imprecision."

"A superintellect has monkeyed with physics."

"It is no more likely that our world has evolved out of chaos than that a hurricane, blowing through a junk yard, should create a Boeing."

"There are many ways of knocking electrons out of atoms. The simplest is to rub two surfaces together."

Fred Hoyle was adamant about his view of the Universe and perhaps pretty disparaging in his remarks. At one of the famed Observatory Club tea meetings, Fred once started a talk by saying, "Oh, Ooh, basically a star is a pretty simple thing." From the back of the room R. O. Redman, a professor of Astrophysics at the University of Cambridge, shouted, "Well, Fred, you'd look pretty simple too, from ten parsecs!"

During the time of his relativity dissertations, Albert Einstein essentially believed that the Universe stood still; that its large scale properties did not vary with time. Most people believe that the Steady-State Theory is one of invariance; that the Universe always existed and that the stars were fixed in place. That is not quite true. Within the framework of relativity, Fred Hoyle developed mathematics making the expansion of the Universe and the creation of matter interdependent. As new galaxies form, matter is created and the Universe can expand. There is no beginning of time. Only "things are the way they are because they were the way they were." The Universe on average is homogeneous and isotropic, of invariant size, in space.

The criticism of the Steady-State Theory is the same as the Big Bang Theory but for different reasons. Radio wave telescopes discovered background radiation that is uniform throughout the Universe. Steady-State Theory cannot explain the uniform background radiation given the deep space between galaxies. The radiation should be less dense between galaxies. There is no explanation for quasars in that their appearance is not uniform throughout space. The Universe is flat and steady-state creation of atoms should appear more three dimensional. In addition as the red shift data began to show galaxies traveling at near the speed of light, the conventional wisdom of a controlled expansion of the Universe became too difficult to justify with a homogeneous view of the Universe.

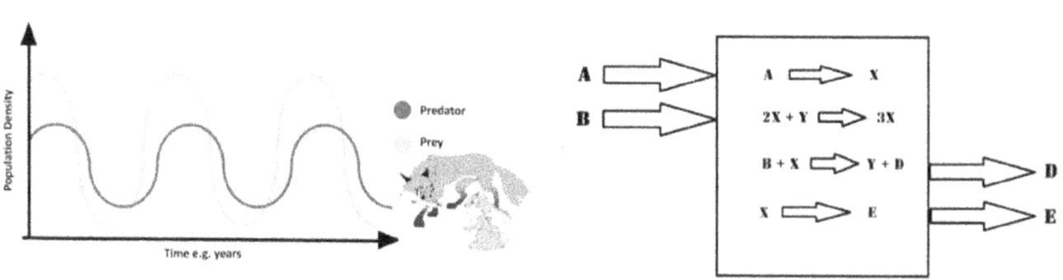

Figure 47: Rabbits Verses Foxes   *Figure 48: Busselator Diagram*

In the mid-sixties, Fred Hoyle gave up. He honestly could not argue against the new findings, particularly from radio telemetry. However, in the early 1980's, the physicist Paul LaViolette proposed his subquantum kinetics theory based on gravity or potential energy wells. He proposed that the spontaneous creation of particles can happen with the minimum energy transfer by the interaction of particle waves. His many books used the well-established Brusselator feedback loop (Figure 48) to describe the process, but it is more easily explained by the logistic equation waveforms (Figure 47) for rabbits and foxes. As the rabbit population increases, the foxes have a food supply. As the rabbit population decreases, the foxes run out of food and their population decreases thus sparking another round of rabbit population growth.

$$(70) \qquad F = R(1 - R) \ where \ F = foxes \ and \ R = rabbits$$

This never ending dance between two particle waveforms can spark a union that can spontaneously generate new atoms. For example, if a photon of light excites an electron, the energy transfer allows the electron to jump to a new orbit. Any excitation can spark a new output (Z) from the Brusselator diagram.

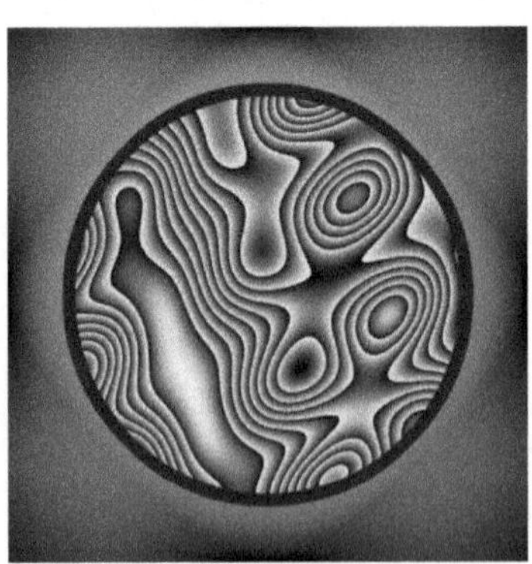

Figure 49: Belousov-Zhabotinskii Chemical Reaction

In 1958, The Belousov-Zhabotinskii experiment showed that a liquid chemical reaction within an enclosed space could spontaneously generate well-ordered waveforms. These slowly moving chemical waves or "reaction-diffusion" waves appear after a dye indicator is added to the reacting solution.

Potentially, LaViolette's hypotheses could spark a new round of steady-state speculation. Continous generation of particles using subquantum energy is easier to digest than the "infinite" energy of the Big Bang. However, LaViolette ties many of his theories to ancient mythology and, therefore, does not have large support among theoretical physicists. For example, in his 1997 book Earth under Fire, he writes about astrology's ether physics as an example of ancient knowledge:

"Gemini (The Twins): Due to the statistical nature of etheron reaction and diffusion processes, the concentrations of these ethers continuously and chaotically pulsate. Dualistic fluctuations emerge in the X and Y ethers (highX/lowY, lowX/highY), each combating its noisy competitors."

"Cancer (The Crab): An emerging (X/Y) fluctuation is protected from the surrounding chaos and amplified by a positive feedback loop in the underlying ether reaction network."

Unfortunately, deep space does not seem to support spontaneous generation as it is largely empty. Also the Universe is flat so that spontaneous generation does not happen in all directions.

3. Fractal Generation

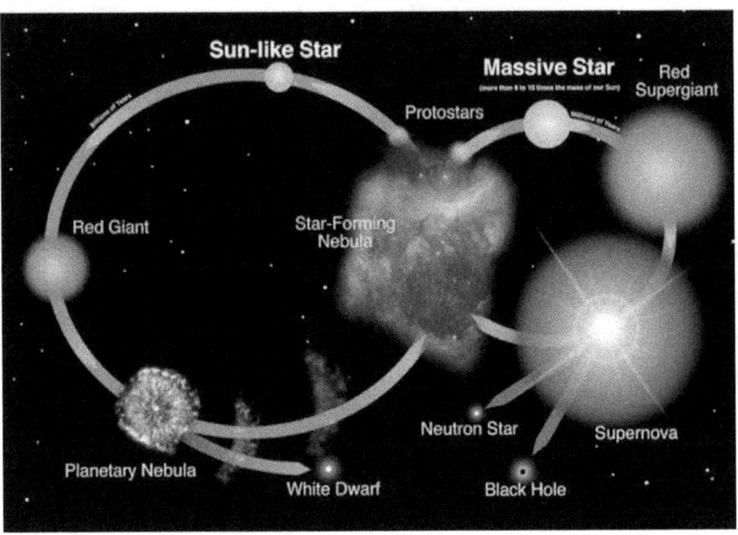

Figure 50: Life Cycle of Stars

Cosmology explains that clouds of nuclear matter ebb and flow to form constellations, suns, solar systems, galaxies, and superclusters of galaxies, and populate space with trillions of stars. This growth from the atom to the galaxy is a self-similar fractal structure. It is a nested hierarchical structure. It is a question of scale.

Robert L. Olbershaw from the Geology Department of Amherst College wrote an essay entitled "A Fractal Universe?" He explains how fractal scaling conforms to the growing fractal structures:

"Given the strong stratification of the cosmological hierarchy, one at least has a very rudimentary self-similar organization: galactic "particles" are composed of stellar "particles", which are composed of atomic "particles." However, this essay proposes that there are more meaningful scaling relations among these "particles" that populate such enormously different size scales. Atoms have radii on the order of 10^{-8} cm while the typical radius for a star would be 5^{1009} cm. So we adopt 5^{1017} as our tentative scaling factor for K lengths, and call it K. Since space and time are treated nearly equivalently in relativistic physics, it will be assumed tentatively that is also the appropriate scale factor for temporal "lengths."

"A promising place to begin the search for self-similarity would be with the well-known analogy between atomic nuclei and neutron stars. In fact the latter have often been likened to "gigantic atomic nuclei." Atomic nuclei typically have radii on the order of 10^{-12} cm and multiplying that value by K gives an estimate on the order of 5 km for neutron star radii. Since neutron stars are believed to have radii of roughly 10 km, the first test of our approximate scaling relation is encouraging. Classical spin periods for atomic nuclei are on the order of 10^{-19} sec to 10^{-20} sec. Scaling these values by K gives expected spin periods for neutron stars on the order of 10^{-2} sec to 10^{-1} sec, which are nicely within the observed range of typical spin periods for neutron stars: 10^{-3} sec to 100 sec. Atomic nuclei also undergo vibrational oscillations with periods of roughly 10^{-22} sec to 10^{-21} sec. Do neutron stars oscillate with periods that are about K times longer? In fact, observed neutron star oscillations are in the range of 10^{-4} sec to 10^{-3} sec, which is in good agreement with our expectations. So for the case of atomic nuclei and neutron stars we find a fairly reasonable degree of self-similarity between the proposed analogues. Both involve extremely dense and rapidly spinning objects whose radii, rotational periods and oscillation periods scale by a factor of K, as expected.

"An important question is whether or not the scale factor K will continue to hold good for extensions to the galactic scale. Galaxies have "peculiar velocities" (random motions in addition to the general Hubble expansion) that average roughly 400 km/sec. Since the galactic "particles" are undergoing such enormously energetic random motions, the only reasonable atomic scale analogues would be fully ionized atomic particles, i.e., nuclei or electrons. Since galaxies are clearly extended objects with considerable substructure, electrons are ruled out as potential analogues. If we scale our nuclear radius (about 10^{-12} cm) by K2 (since we are "going up" two scales), we get a predicted radius on the order of 10^{23} cm for a typical galaxy. Again this is correct; galactic radii typically are on the order of 10^{22} cm to 10^{23} cm. Moreover, if we scale typical nuclear spin periods by K2, we get the correct order of magnitude for galactic spin periods: on the order of 10^8 years."

Since the Universe grows in conformance with fractal self-similarity, the Fractal Theory becomes more of a steady-state theory than an instant creation. Therefore, it suffers the same questions about the Steady-State Theory; namely, background radiation, quasars, the rapidly expanding Universe, and the Universe's relative flatness.

4. Pulse Theory

At an earlier time, Pulse Theory was part of the Big-Bang Theory. It was presumed that Universe expansion would eventually slow down, and the gravity caused by the galaxies created would eventually collapse the Universe. Then the Big-Bang would happen all over again.

Because of the red-shift observations, it is possible that galaxies may be escaping the observable Universe because they are traveling faster than the speed of light. These incredible speeds argue against Pulse Theory, but Pulse Theory is a more reasonable alternative.

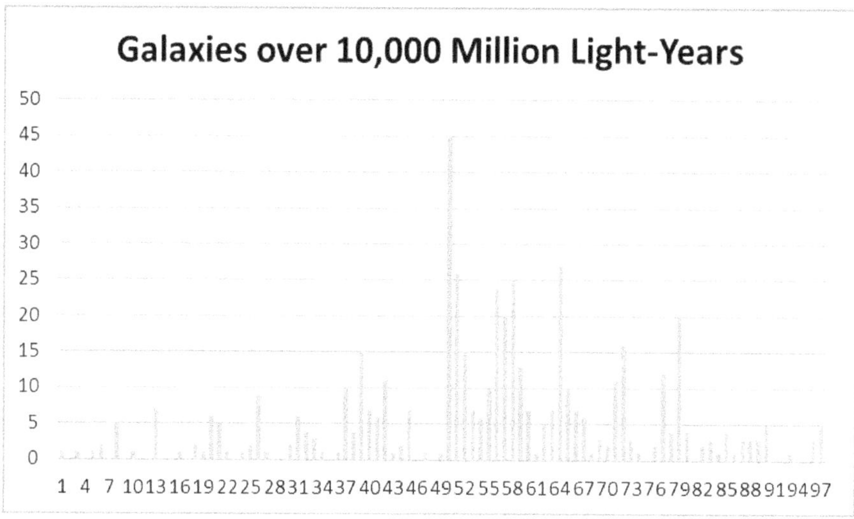

Figure 51: Galaxy Count over 4 Billion Light-Years

The final explanation brings us back to Olbers' Paradox. If in the Big Bang Theory the Universe is expanding, then it seems reasonable to assume areas of darkness. However, the Andromeda Galaxy is moving toward the Milky Way Galaxy. The galaxies will collide in one billion years. Why then does not the Andromeda Galaxy, within Earth's sightlines of three degrees and with one trillion stars, not look like a searchlight? Our closest galaxy should glow. The same for the Steady-State Theory. If most of the time stars are in fixed positions, pocket clusters of stars should at least glow like the Northern Lights. This brings us to Fractal Theory. You can think of the Universe as being smooth because some argue that it is homogeneous. But what if the Universe is rough? Its fractal dimension can be measured.

From Paul LaViolette's 1995 book <u>Beyond the Big Bang</u>, there is a chart on the number of galaxies that have been counted over a 10,000 million light-year period with Earth at its center. The chart has been recreated above. Sometimes there are gaps of approximately 450 million light years between galaxies. The chart yields 97 data locations with magnitude; a two-dimensional array. With this limited data set, an approximation to the fractal dimension can be calculated. If the fractal dimension is near 2, then the Universe is homogeneous and should present a blinding light. If the fractal dimension is near one indicating a very smooth area, then the Universe is nearly empty of stars.

The fractal dimension of the Galaxy Chart is 1.4. This number suggests that much of outer space is dark.

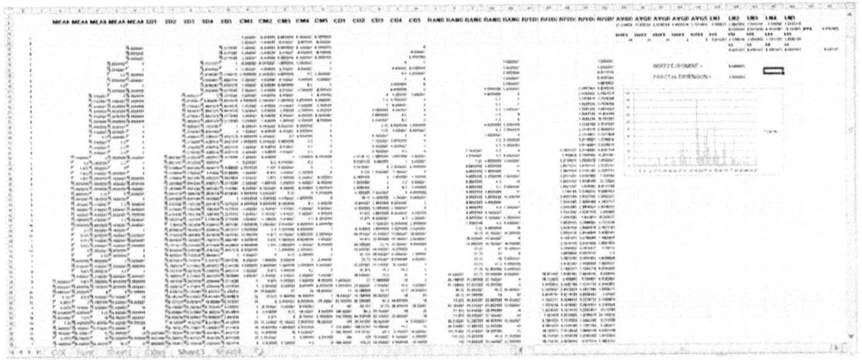

Figure 52: Excel Calculation of the Fractal Density of Stars

Chapter 10: Roughness

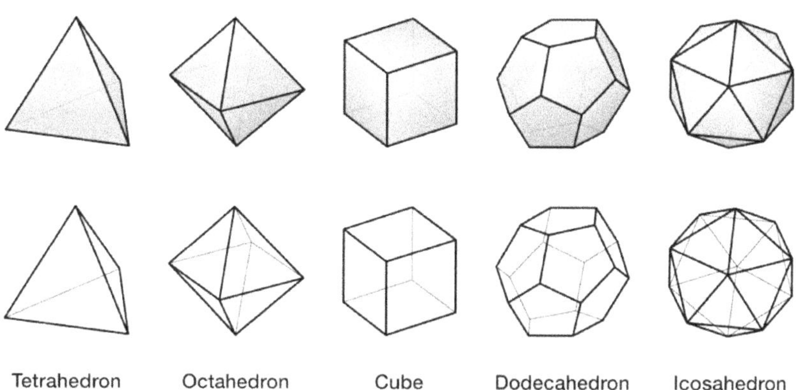

Tetrahedron   Octahedron   Cube   Dodecahedron   Icosahedron

Figure 53: Platonic Solids

When the first Greek philosophers began their scientific investigations, Plato in Timaeus stated that the Universe was built on five "Platonic" solids. The scientific community can confirm that these are the only solid structures with identical sides that can exist in Nature. In fact, the smooth sides of these solids do not exist in Nature. Buildings that look like cubes or cylinders exist but these geometries do not exist in Nature. Nature is rough.

In 1975, the American mathematician Benoit Mandelbrot coined the word "fractal" to describe Nature. Fractal geometry is a new geometry to describe the irregularities of Nature. For example, your lungs are not balloons but have tree branches that are fractal in Nature. Medical research can use that fact to better pinpoint emerging tumors. The tree on the right (next page) is a simple fractal structure. Break the trunk of the tree into two halves and place the branches in a "V" shape on top of the trunk. Break the two branches in half and complete the "V" shape on top of each branch. Repeat again. Keep repeating. The entire scenery for Disney's "The Jungle Book" was artificially created with algorithms like this.

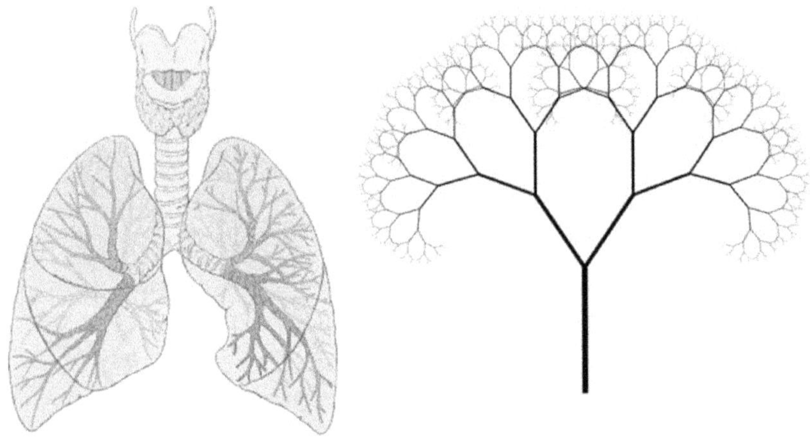

Figure 54: Fractal Nature of Lungs   *Figure 55: Fractal Tree*

The above objects can be described as self-similar. A branch looks like a branch which broken in two still looks like a branch. Recalling the fractal dimension equation:

$$(71) \quad \textbf{\textit{Fractal Dimension}} = \textbf{\textit{FD}} = \frac{\log(\textit{similar objects})}{\log(\text{scalar dimensions})}$$

$$(72) \quad \textbf{\textit{For the Tree}} \quad \textbf{\textit{FD}} = \frac{\log(2)}{\log(2)} = \textbf{1}$$

The result for the tree shows identical similarity among all branches. This result is described as smooth. Certainly the tree is an idealistic realization. If we were to slice the lung into a two dimensional digital image like a catscan or an MRI, the fractal dimension possibly would be between 1.25 and 1.5. Obviously, the lung has much rougher branching. Both structures are examples of the self-similar nature of living fractals.

Another method for calculating a fractal dimension is to overlay a graph paper grid on top of the two dimensional structure that you are measuring. The coastline of Britain can be measured this way, but you don't have to put a piece of paper over the entire coastline. A photograph or a scaled drawing of the coastline will do. The drawing below overlays the outline of the coastline on three different box sizes of graph paper. The boxes that touch the outline are counted. Then three points on a log-log graph are plotted. The slope of the line is the fractal dimension.

Figure 56: Fractal Dimension of British Coastline by Box Count

$$(73) \qquad FD = average\left(\frac{LN(132)-LN(53)}{LN(16)-LN(4)} + \frac{LN(53)-LN(22)}{LN(4)-LN(1)}\right) = 1.36$$

The dimension can also be found mathematically by calculating the average slope:

No. Squares	LN Squares	Normal Scale	Inverse Scale	LN Inv. Scale
132	4.883	1	16	2.773
53	3.971	4	4	1.386
22	3.091	16	1	0

The calculation compares with the published calculation for the coast of Britain of 1.21. The formal calculation found more data points by using a lot more graph paper.

Note that the coast of Britain is not an example of self-similarity. Chaos Theory is not just about the recursiveness of Nature. It is about the roughness of Nature and how to model its properties.

This method can be used to help diagnose medical problems. By using multiple computer-generated grids, the fractal dimension of a two-dimensional lung image can be calculated. If there is an obstruction such as the tumor pictured below, then the fractal dimensions should increase because of the increasing roughness. This process could be used to catch tumors earlier that are hard to detect visually.

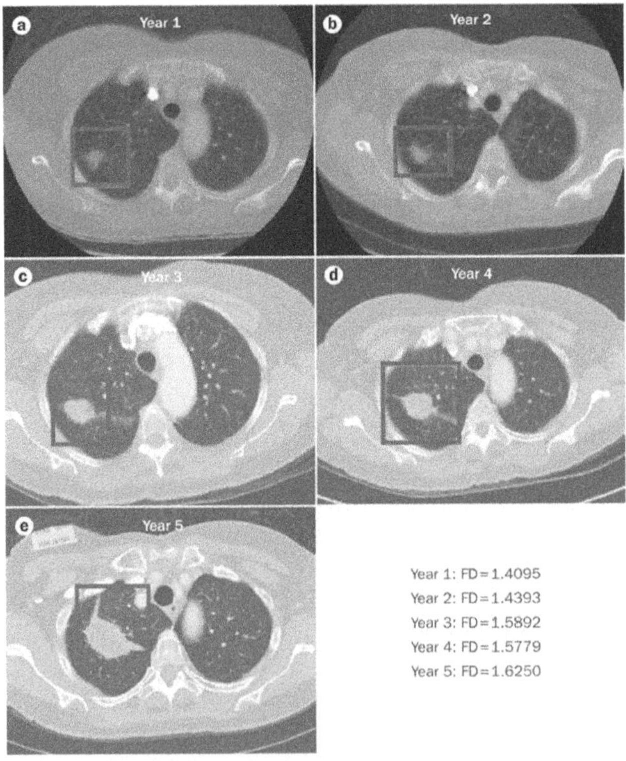

Figure 57: Fractal Diagnosis of Advancing Lung Cancer

Mathematical differential equations basically used to model dynamic man-made objects have progressed in recent history to model difficult to solve equations by graphical methods. Normally these methods lead to attractors or repellers that are labeled sources or sinks. Dynamic Chaos Theory that models living systems has stranger properties.

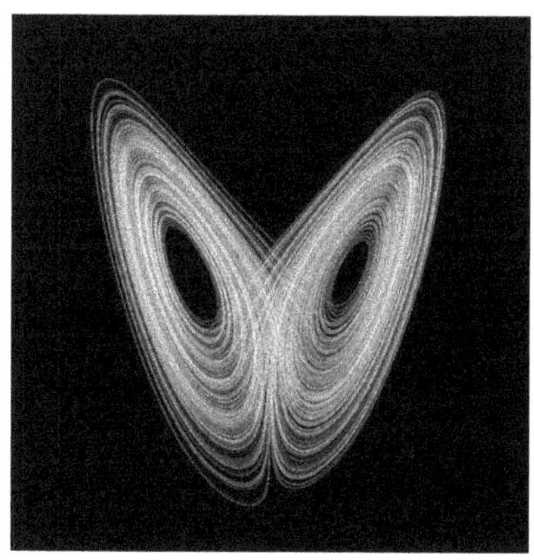

Figure 58: Butterfly Effect

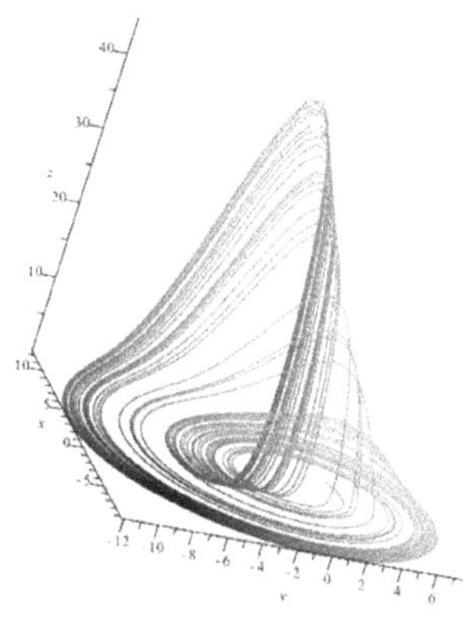

Figure 59: Rossler Effect

Edward Lorenz (1917 – 2008), mechanical engineer and mathematician, developed at MIT a complex series of differential equations to model weather patterns in the early to middle 1960s. The computers in those days were not much help in obtaining results. Frustrated, he simplified the equations and then simplified them again. The result was the upper picture on the left in the shape of a butterfly that became known as the "butterfly effect". The concept of a butterfly flapping its wings and causing a repetitive growth of wind strong enough to knock over a mountain became part of the folklore of science. The public liked it. What Lorenz really invented was Chaos Theory; a way to represent dynamical systems in Nature. Next are his simplified equations for the butterfly effect.

$$(74) \qquad \frac{\partial x}{\partial t} = \sigma(y - x)$$

$$(75) \qquad \frac{\partial y}{\partial t} = x(p - z) - y$$

$$(76) \qquad \frac{\partial z}{\partial t} = xy - \beta z$$

What these equations represent is something very special. First, the initial conditions σ, ρ, and β are described as unstable. The size and shape of the butterfly cannot be predicted. Then, the orbit lines in three-dimensional space following another line cannot be predicted. All that can be said is that they will orbit around to a zero point between the wings. They will not intersect the zero point. In fact, an orbit will not intersect (cross) another orbit. There will always be space between orbits. The "zero point" is called a "strange attractor". It is strange in that, unlike a sink in differential equations where all solutions within a certain region intersect the sink, no orbital solution intersects the strange attractor. With all this chaos, the fact that the solution is stable, that a butterfly form results, is one of the most important discoveries of science.

The randomness of the orbits resulting in a stable solution sounds like the number game that we played previously. Given a process, in this case three equations, the result of a chaotic process is a stable solution.

This phenomena is better explained by the Rossler effect (Figure 59), another set of three differential equations. As shown, think of multiple orbits as ribbons of clay. Note that there is space between the ribbons; they do not intersect. In fact, each orbit is like folding filo dough for baking. You have a pile of dough that you flatten and stretch. Then you fold the dough to form two layers. Again you flatten, stretch, and fold over. Now you have four layers. The process is repeated and now you have eight layers. This is a bifurcation process. With each orbit ribbon at the strange attractor, you are doubling the layers. There is space between the layers. The layers do not touch. When you eat a filo dough pastry, you are not eating one solid lump of crust. You are eating layers of pastry.

Perhaps the most famous images in chaos theory are the Julia Set (lower left) and the Mandelbrot Diagram (lower right). There are many ways to create a Julia set by reiterating a repetitive function to create his beautiful images. Gaston Maurice Julia (1893 – 1978), a French mathematician, won the Grand Prix from the French Academy of Science in 1918 for his paper on the iteration of polynomial functions. He proved that the functions are infinite and self-similar at various scales. With his definition of fractals in 1975, Benoit Mandelbrot (1924 – 2010) developed his Mandelbrot set that goes a long way in defining the properties of this geometry.

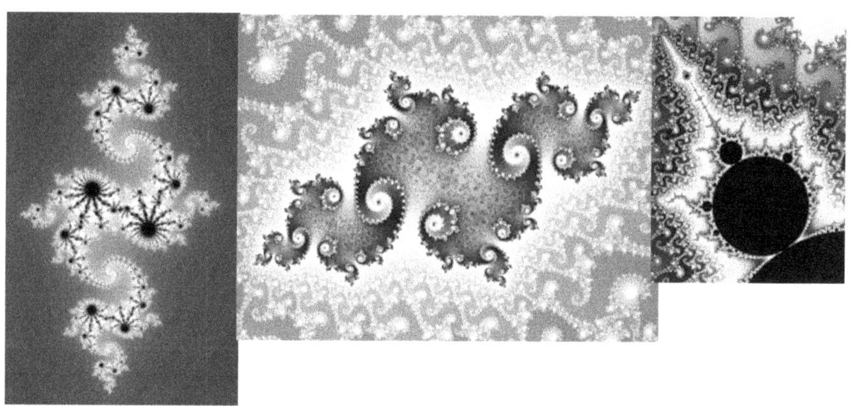

Figure 60: Julia Set and Mandelbrot Diagram

Every part of the Mandelbrot diagram is a scaled self-similar part of another part of the diagram. Just like the Lorenz diagram or the butterfly diagram, you can think of it as a periodic recursiveness with period p. The Mandelbrot set has many properties, most notably about the increasing complexity of the small nodules in a counterclockwise direction, but the properties that are of most interest are the lengths of the orbits. The butterfly and the Rossler effects contain unequal orbits. This leads us to Sharkovsky's Theorem.

All iterative functions may intersect the 45 degree "real line" in two-dimensional space. Those intersections are fixed points. These points may be (strange) attractors, repellers, or neutral in that the point attracts on one side and repels on the other. If there are two fixed points, then this describes a bifurcation of the iterative 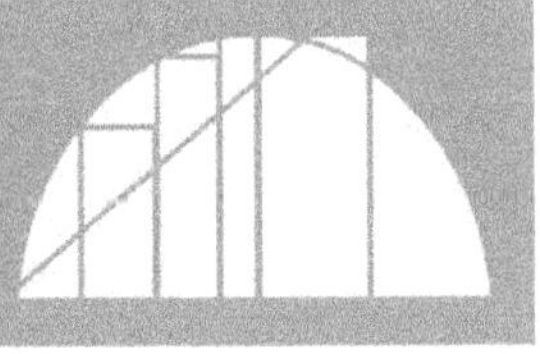

Figure 61: Sharkovsky Real Line

function. However, as the iteration continues the orbits may be unpredictable or chaotic. There may be 5 intersections followed by 3 intersections and so forth. These intersections are bifurcations but they do not have to be in order of increasing magnitude.

Sharkovsky had an ordering of the fixed points:

3,5,7,9,11, …
2x3, 2x5, 2x7, 2x9, 2x11, …
4x3, 4x5, 4x7, 4x9, 4x11, …
8x3, 8x5, 8x7, 8x9, 8x11, … 16x3, …
2, 4, 8, 16, …, 1

Oleksandr Sharkovsky, a Ukrainian mathematician, developed his discrete periods for dynamical systems in 1964. The Theorem is as follows:

> Assume that f is a continuous orbital map for an interval and a prime orbital period p of f such that p < q, then f has a periodic orbit containing prime period q.

What this means is if there were 12 fixed points then there must exist bifurcations in the table for numbers to the right of 12 but not for the numbers or lines to the left of 12. If you can find 3 fixed points, then all bifurcations in the table for the iterative function f exist.

The main idea is that orbits are whole numbers but not decimal numbers or irrational numbers. This next orbital map may be the most important in the book.

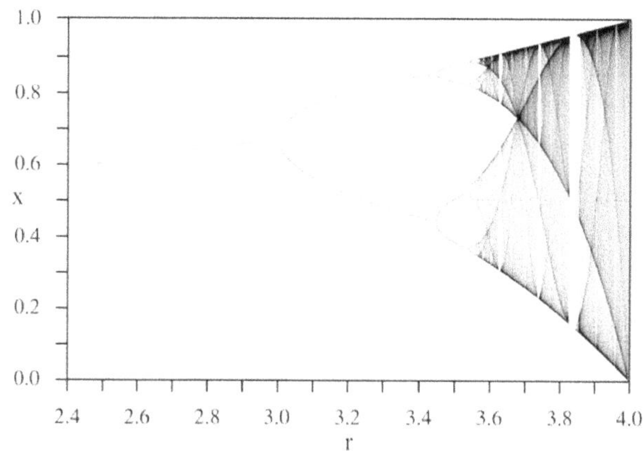

Figure 62: Bifurcation Map

The map is for the logistics equation of rabbits and foxes. The first node shows a 2 path bifurcation followed by a 4 path bifurcation. Those strips or areas of white are stability areas where bifurcations or folding and stretching for the Rossler equations take place. The areas of gray are chaotic areas where nothing exists such as the space between folds where nothing survives. For example, sound waves in an enclosed room are either reinforced or cancelled out. A sin wave interacting with a cosine wave of the same magnitude cancels to zero. A new wave interacting in a gray area will randomly be cancelled out due to the chaotic noise of many frequencies. However, if the new wave is of a like period in a stability area, it will be reinforced with the current iterative equations.

If this book contained a dramatic play, then the gray areas would be orbits or periods of death, and the white areas would be the orbits of life. The drama of this play is this subject of this book.

Chapter 11: Telegrapher's Thermodynamics

What does the Telegrapher's Equation have to do with thermodynamics? Well there is a delivery or dissipation part and there is a dispersion or diffusion part. The diffusion part is thermodynamics.

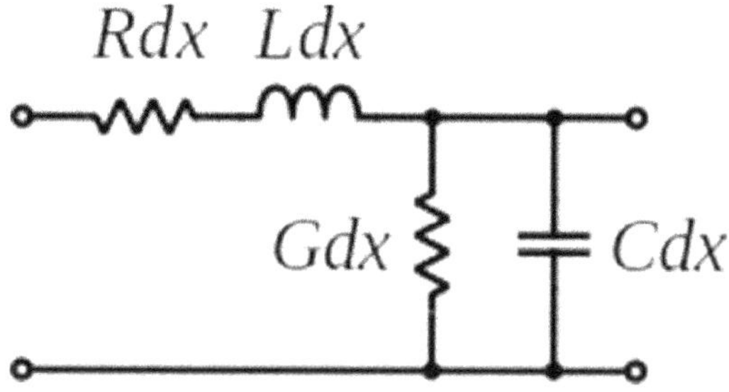

Figure 63: Telegraph Circuit Schematic

There are two parts to a transmission or telegrapher's line. The transmission line delivers current to a number of nodes such as streetlights or a neighborhood community of homes or to the night banks of lights in a football stadium. The inductor L traps excess current between nodes and releases the current when demand is heavy. The box underneath the transmission line represents the diffusion outlet to the station being powered. The tapped current is conducted where needed. The capacitor C blocks excess voltage and drains the voltage to ground when not needed. Power lines in the United States operate at 120 volts and 240 volts.

$$(77) \qquad \frac{\partial^2 v}{\partial t^2} + a\frac{\partial v}{\partial t} + bc = c^2 \frac{\partial^2 v}{\partial x^2} \qquad \text{where}$$

- $c = \frac{1}{\sqrt{LC}}$ *a frequency*
- $a = c^2 (LG + RC)$ *a nodal resistance sum for current*
- $b = c^2 RG$ *a voltage resistance drop for diffusion*
- $R = resistance$
- $G = conductance = \frac{1}{R'}$
- $i = current$
- $v = voltage = iR$
- $L = inductance = v\frac{\partial t}{\partial i}$
- $C = capacitance = i\frac{\partial t}{\partial v}$

Because inductance and capacitance have derivatives with respect to voltage and current, their values change with respect to time. In other words, the transmission line is a waveform. Resistance and conductive values are linear scalars. Given that the diffusion circuit is detached from the transmission line, it will simply die. New York City will have a blackout. The diffusion circuit could be a model for the Second Law of Thermodynamics.

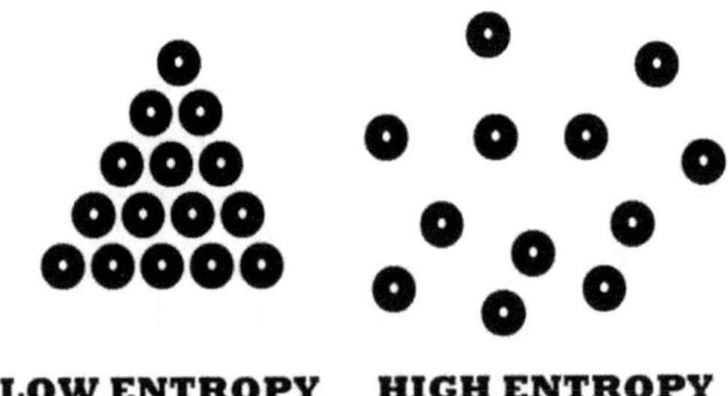

Figure 64: Particle Entropy

The Second Law of Thermodynamics can be stated in many different ways. Essentially, it is a law of disorder. If you crack an egg, it cannot be put back together again. As entropy increases, disorder increases. Once an entity loses its source of energy, it falls to its lowest state. This might be from a liquid state to a crystalline solid. In outer space, a gas will fall from its normal temperature to the temperature of outer space at 2 degrees Kelvin. As a gas falls in temperature, its molecules drift farther apart. Without a new source of energy, they can never raise their temperature; they can never come closer together. Entropy is irreversible. It can never decrease; it can only increase. The entity is in a diffusion state.

More formally, The Second Law of Thermodynamics states that temperature differences between systems in contact with each other tend to even out. Work can be obtained from these non-equilibrium differences, but loss of thermal energy to do work means that entropy will increase.

$$(78) \qquad \partial S = \frac{\partial Q}{T} \qquad where$$

- $S = entropy\ in\ joules\ per\ Kelvin$
- $Q = heat\ energy$
- $T = common\ temperature$

A common misconception is that entropy cannot decrease tending toward order. In general for a hot liquid mixing with a cool liquid, the temperature will average out. The hot liquid will increase its entropy as the hot liquid cools and its

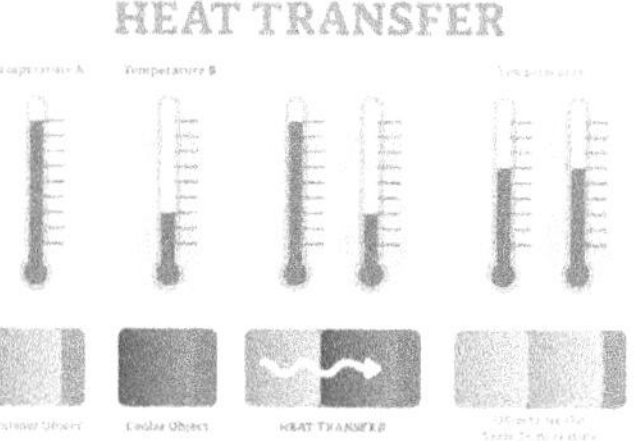

Figure 65: Heat Transfer

molecules separate. However, the temperature loss equals the temperature gained for the cool liquid. The conservation of energy is the First Law of Thermodynamics. The entropy for the cool liquid has decreased and the liquid has become more organized.

Scientists say that the decrease in entropy does not violate the Second Law of Thermodynamics. In a local environment the energy is transferred and the total energy is retained. This violates the narrow intention of the Second Law. The Second Law is one of decay. It is thought of as a death spiral or at least a death curve.

Let us assume quantum particle random motion; the Brownian motion of Einstein's distance law ($d = \sqrt{2} * N\_steps$). Albert Einstein explained that the Brownian motion of a particle with N-steps has a boundary of plus or minus $\sqrt{2} * N\_steps$. The Martingale process of this motion means that the expected value after N-steps is 0 (zero). In this general case, the average value is the expected value. Note also that the Hurst exponent of any random process is 0.5. Also, note that a random process has no memory.

Suppose on the X-axis we put the black dot at 0 (zero) and then let it take N steps (where N is any number). Now we want to know how far the black dot travels after it has taken N steps. Of course the distance traveled after N steps will vary each time we repeat the experiment, so what we want to know is, if we repeat the experiment many, many times, how far the black dot will have traveled on average. Let's call the distance that the black dot has traveled "d". Keep in mind that d can either be positive or negative, depending on whether the black dot ends up to the right or left of 0. We know that for any one time that we repeat the experiment,

$$(79) \qquad d = a_1 + a_2 + a_3 + \ldots + a_N$$

Now we use the notation $<d>$ to mean "the average of d if we repeated the experiments many times."

$$(80) \qquad <d> \leq (a_1 + a_2 + a_3 + \ldots + a_N) \geq <a_1> + <a_2> + <a_3> + \ldots + <a_N>$$

But $<a_1> = 0$, because if we repeated the experiment many, many times, and a_1 has an equal probability of being -1 or +1, we expect that the average of a_1 will be 0. So then,

$$(81) \quad <d> = <a_1> + <a_2> + <a_3> + \ldots + <a_N> = 0 + 0 + 0 + \ldots + 0 = 0$$

This isn't too surprising if you think about it. After all, $<d>$ is the average location of the black dot after N steps, and since the dot is equally likely to move forward or backwards, we expect d to be 0, on average.

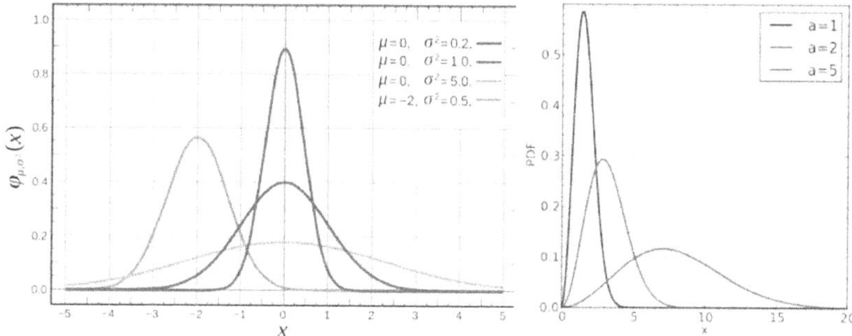

Figure 66: Normal Distribution   *Figure 67: Maxwell-Boltzmann Distribution*

The random walk with particles in space fits in with the notion that particles do not collide. This perfectly fits the random nature of a Normal Distribution. There will be no temperature change because, in the land of chaos, nothing happened. However, what if they do collide? If the particles travel fast enough they may excite other particles either to jump to another orbit or leave their atom altogether and become ionized. If that happens both photons can be released weakening the energy level of an atom and ionized particles may bond with other particles forming new atoms or molecules. One thing is for sure; it will become hotter and more ordered. The Maxwell-Boltzmann Distribution models colliding particles within a closed space. Both distributions are bell-shaped curves, but the Maxwell Distribution has a tail. At the end of the distribution, particles are traveling faster, and with an excitation level, the tail gets pushed out. Clark Maxwell, the father of electromagnetics, proved that particles in a box that collide with one another follow his distribution. The Maxwell-Boltzmann Distribution is a velocity (speed) distribution.

This is where things get messy. We have to make some assumptions in order to compare the two distributions. First, we must convert the Normal Distribution into a velocity (v) distribution. Let us define two velocity density distributions: D_N for the Normal Velocity Distribution and D_M for the Maxwell-Boltzmann Distribution. We must add velocity terms to the Normal Distribution. Both distributions are probability density distributions whose area under the bell shaped curves is equal to 1.

$$(82) \qquad D_N(v) = \frac{1}{\sqrt{2\pi\sigma^2}}\, v^2\, e^{-\frac{v^2}{2\sigma^2}} \quad where$$

- μ = the mean or expectation of the distribution (and also its median) = 0
- σ = the standard deviation
- σ^2 = the variance

$$(83) \qquad D_M(v) = \sqrt[2]{\frac{m}{2\pi kT}}^{\frac{3}{2}} 4\pi v^2\, e^{-\frac{mv^2}{2kT}} \quad where$$

- $m = mass$
- $k = Boltzmann's\ constant$
- $T = temperature$

While these equations are not solved easily without a table of results, most of each equation is filled with constant values. Let us construct a general equation. For the constants,

$$\text{(84)} \qquad C_{1N} = \frac{1}{\sqrt{2\pi\sigma^2}}$$

$$\text{(85)} \qquad C_{2N} = \frac{1}{2\sigma^2}$$

$$\text{(86)} \qquad C_{1M} = 4\pi^{\frac{3}{2}} \sqrt{\frac{m}{2\pi kT}}$$

$$\text{(87)} \qquad C_{2M} = \frac{m}{2kT}$$

Now both equations have the following general form:

$$\text{(88)} \qquad D(v) = C_1\, v^2\, e^{-C_2 v^2}$$

By taking the derivative and setting it equal to zero, the equation can be solved for the most probable velocity of each distribution. The most probable velocity value, meaning the most particles with that velocity (maximum value), will be the highest value for each distribution.

$$\text{(89)} \quad \frac{\partial D}{\partial v} = C_1\, e^{-C_2 v^2} * (2v) + C_1\, v^2 e^{-C_2 v^2} * (-2C_2 v) = 0$$

$$\text{(90)} \quad \frac{\partial D}{\partial v} = 2C_1 v e^{-C_2 v^2} - 2C_1 C_2 v^3\, e^{-C_2 v^2} = 0$$

$$\text{(91)} \quad \frac{\partial D}{\partial v} = 2C_1 v e^{-C_2 v^2} (1 - C_2 v^2) = 0$$

$$\text{(92)} \quad v = \sqrt{\frac{1}{C_2}}$$

Each distribution has a C_2 that can be equated to most probable velocity:

$$(93) \quad v_N = \sqrt{2\,\sigma^2} = \sqrt{2}\,\sigma = average\ velocity$$

$$(94) \quad v_M = \sqrt{\frac{2kT}{m}} =$$

most probable velocity

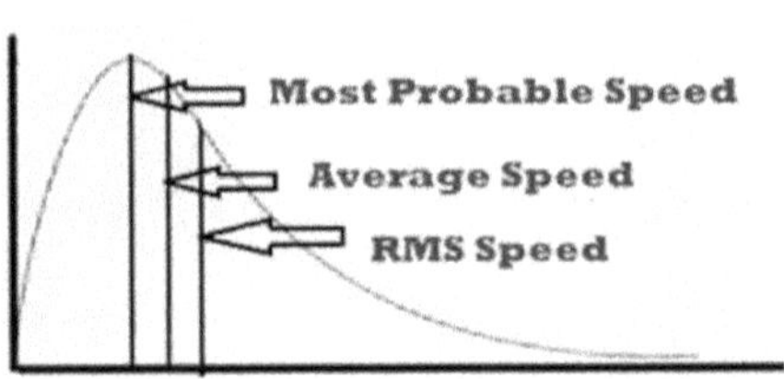

Figure 68: Maxwell-Boltzmann Speed Averages

For a distribution that is not symmetric, the most probable velocity is not equal to the mean or average velocity. Using substitutions similar to the math for the most probable velocity, the average velocity $(avg\ v_M)$ for the Maxwell-Boltzmnn Distribution can be calculated. Given the same conditions – number of particles, temperature, and the same enclosed flask – the average velocity for Maxwell should equal the mean velocity for the Normal Distribution.

$$(95) \quad avg\ v_m = \sqrt{\frac{8kT}{\pi m}} \cong v_N > v_m$$

$$(96) \quad \sigma_N = rms\ v_N = \sqrt{\frac{8kT}{\pi m}}$$

Given that there are no collisions between particles, the Maxwell Distribution will look like the Normal Distribution. The temperature of the light gases would have to be at least 2,000 degrees Kelvin for an excitation level to appear and the possibility of chemical bonding to occur. When this happens, the most probable velocity v_M shifts left from the mean velocity v_N of the Normal Distribution. At the same time, the Maxwell Distribution flattens and extends, its value $(rms\ v_M)$ becomes greater than $(rms\ v_N)$.

Speed is an indication of temperature (Equation 95). Speed can also be expressed as:

$$(97) \qquad rms\ v_M = \sqrt{\frac{3kT}{m}} = \sqrt{\frac{3RT}{M}} \qquad where$$

- $R = 3.814 = ideal\ gas\ constant\ in$
  $joules\ per\ mole - degree\ Kelvin$
- $M = atomic\ weight\ in\ grams\ per\ mole = 4.002\ for\ He$

Assuming that the gas helium (He) is heated in a container to 2,000 degrees Kelvin, an excitation level can be estimated from Equation 97.

$$(98) \qquad excitation\ v_a = \sqrt{\frac{3*3.814*2000}{4.002}} = 302\ meters/sec$$

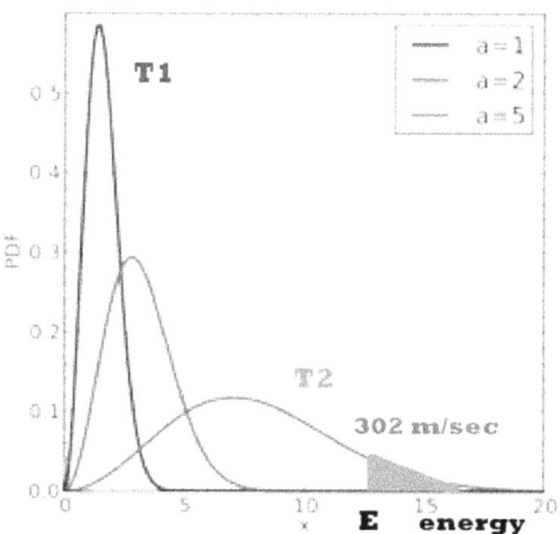

Figure 69: Reverse Entropy Distribution

This graph begs the question: Given an increase in temperature from T1 to T2 caused by the excitation of molecules, should the Laws of Thermodynamics model a reverse entropy process whereby energy is infused into a process and entropy is decreased? Also, should the law be for a new reverse entropy process be given a name for a field induction that measures order from zero degrees Kelvin in joules per Kelvin?

Corollary to the Second Law of Thermodynamics

The "intropy" P of an isolated system infused with a field induction process however temporary, will dynamically increase in both temperature and potential energy, because isolated systems that spontaneously evolve move towards a higher thermodynamic equilibrium—the state of maximum intropy. Intropy is defined as a measure of potential energy in joules per Kelvin as normalized from zero degrees Kelvin. As long as the process remains dynamic in a steady state sense or is infused (grows) with new energy, then the intropy will not decrease but can only increase.

$$(99) \qquad \Delta P = \frac{1}{2}m(\Delta v)^2 = \frac{1}{2}m * \frac{2k\Delta T}{m} = k\Delta T \qquad \text{where}$$

- $k = Boltzmann's\ constant =$
  $1.380649 \times 10^{-23}\ joules\ per\ Kelvin$
- $T = temperature\ in\ degrees\ Kelvin$
- $\Delta T = \frac{T_2 - T_1}{T_1}\quad (dimensionless\ and\ T_2 > T_1)$

$$(100) \qquad P = kT$$

The word extension "tropy" means a change in response to a stimulus. In an engineering sense, entropy means decay and intropy means gain. In a general sense, entropy means death and intropy means life.

Chapter 12: Bandpass Filter

Figure 70: Battle of Britain Memorial - St. Paul's Cathedral Still Stands

In July 1940, the Battle of Britain began. Germany would fly across the channel to bomb the city of London. It was a historical fact that St Paul`s Cathedral, the symbol of resistance during the Blitz, remained standing while all around was demolished. The problem was that this was the beginning of the war for Britain and they did not have enough fighter planes to protect the entire coast. Radar could detect the planes and the British force could be re-positioned to meet the enemy. Germany countered by zig-zagging their flight patterns so that the British would not know where the Nazi planes would cross the coast.

A statistician was brought in to examine the German

tendencies. He recommended that when radar sightings were detected, all data should be averaged. This simple mathematical trick pinpointed where the Germans would cross the channel. The weaker forces of Britain could be deployed tactically. Because Britain would not surrender despite the raids, the Battle of Britain was considered a victory.

There is another way to filter out noise; the bandpass filter. A bandpass filter sets a range of frequencies and ignores all frequencies above and below the range. For example, given an audio file with static-type noise, you might want to filter out everything but the range of the human voice from 100 hertz to 500 hertz to listen to a conversation.

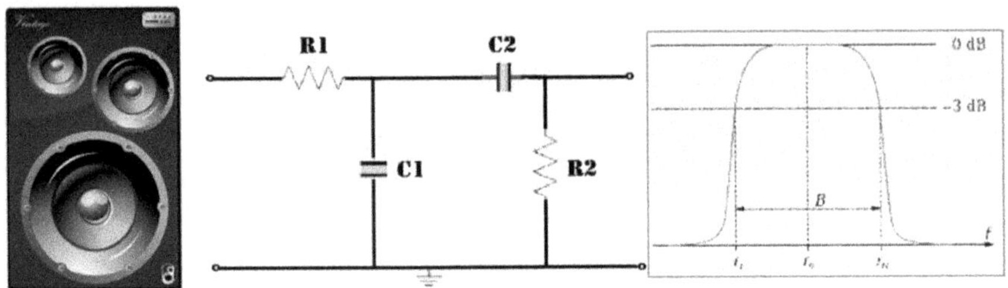

Figure 71: Bandpass Filter Circuitry for Speakers

In the days of component stereo systems before the advent of the cell phone, loudspeakers typically contained a tweeter, a mid-range, and a bass cone. These speakers typically pulsed as they vibrated, and if the speaker had to handle too many frequency pitches, say a full organ, then their oval geometries would distort the sound. The idea was to find the right size speaker cone and limit the frequencies that the speaker cone could handle; hence, the bandpass filter circuit. A bandpass filter consists of a low pass filter and a high pass filter. A filter circuit consists of a resistor R and a capacitor C . Their values can limit the frequency waves that enter a cone. As the plot shows, the frequencies will steeply roll off at the bandwidth edges depending on how many R + C combinations are used. As the plot shows, these combinations cause a phase delay. The greater the delay, the more likely the sound image will not sound sharp or focused. Higher sounds are easier to pinpoint in a stereo image than lower sounds. Typically a bass cone will support a frequency range from 20 to 125 hertz, a mid-range from 125 to 1,000 hertz, and a tweeter from 1000 to 20,000 hertz.

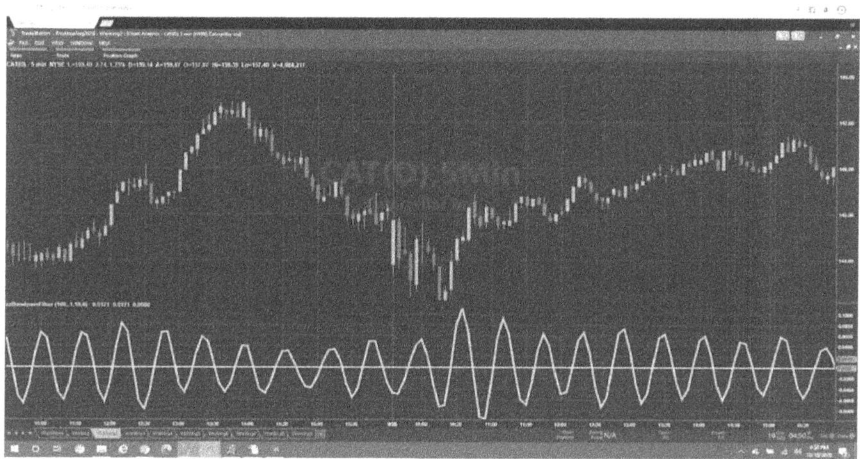

Figure 72: Bandpass Filter of Stock Noise

If you trade stocks, the concept of noise in the activity of a stock might strike a trader as not real. After all, buyers equal sellers. Yet the above chart shows a filtered waveform of the day-trade activity of Caterpillar (CAT).

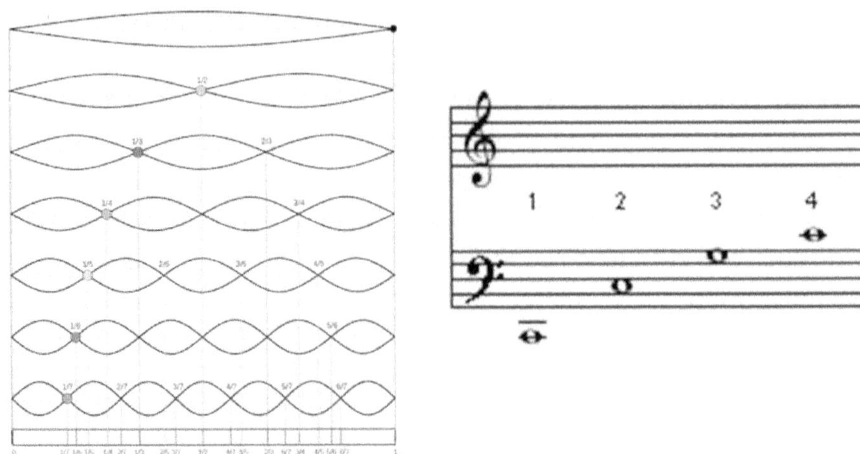

Figure 73: Stringed Instrument Waveforms

You might recall the overtone series of musical notes from Chapter 6 "Partial Order". From a low C, the first overtone has a 2 to 1 ratio to the fundamental tone and is also a C-note an octave above at twice the frequency. The next overtone G is 3 to 2 times the previous frequency. The fourth tone is 4 to 3 times the previous frequency. These tones can be reproduced on a string instrument given the right length of string for a low C. By plucking the string you hear the note. By placing your finger one-half way (1 to 2 divisions) you hear the first overtone. By placing your finger one third way (2 to 3 divisions), the second overtone G is heard. Similarly, 3 to 4 divisions on a plucked string yields the third overtone, two octaves higher and 4 times the frequency of low C.

What if we applied this logic to a "noisy" chart in the stock market? A caveat: this is not financial advice. The author is not a financial advisor; nor is he trading stocks at the time of this book. Lots of stock market analysts can give you technical trading advice on cycle trading.

Let us start again. Consider an objective of analyzing a stock market chart on a day-trading 10 minute scale. Generally, a trader might be interested in a trend. Is the stock trading up or down? The problem is: If you are trading up and the stock market at a higher scale - say 20 minutes - is trading down then your trade over enough time will likely lose. You might want to trade up given your plot patterns or analysis only when the stock at a higher scale is trading up, and trade down with your analysis only when the stock at a higher scale is trading down.

The proposition becomes: Let us define a fundamental frequency or cycle at a higher scale and sift downward in scale at faster frequencies just like the plucked strings of music. Let us look at a 40 minute scale, then a 20 minute scale, a 13 minute scale, and finally, the 10 minute scale of our hypothetical trade analysis. In the order of a plucked string, the scale ratios are 1 to 1, 1 to 2, 2 to 3, and 3 to 4. The 13 minute scale is a round-off of a bar time of 13.33 minutes as the closest the trading platform would allow for minute units. For the stock Honeywell (HON), here are the waveform charts.

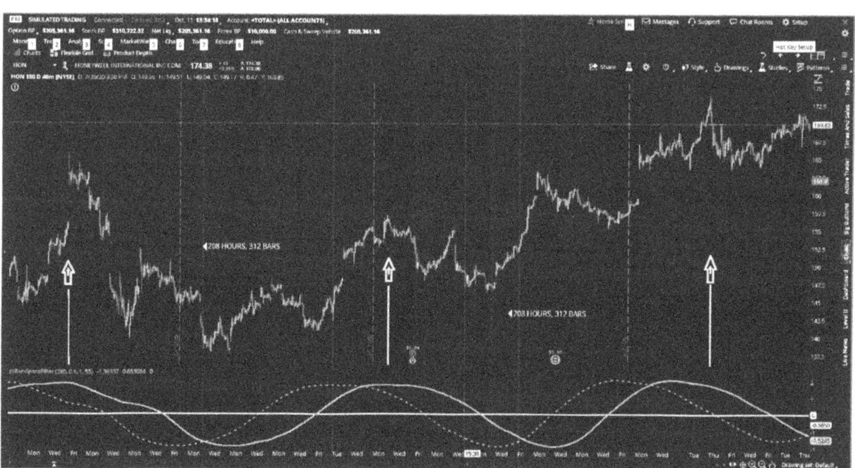

Figure 74: 40 Minute Chart (208 Hours, 312 Bars)

The John Ehlers bandwidth routine is sophisticated. John Ehlers is an electrical engineer and a noted author on methods of stock trading. The routine takes into account a limited bar bandwidth and the number of bars considered, and then converts the angle of the last three bars and the estimated phase delay into a waveform. The waveform can be "tuned"; in this case to 280 bars. The chart shows a waveform aligned with three significant peaks. The distance between these peaks becomes the bar length of the musical string and the chosen fundamental frequency of the plot. Given only the Monday through Friday bars for the 6 ½ hour trading day on the New York Stock Exchange, and given that weekends and holidays don't count, the string length is 208 hours. The hours are estimated from the dates involved and may be under-counted or over-counted by a weekend of 2 days and the holidays by 1; roughly a 10% error.

The next chart, a 20 minute chart, has a 1 to 2 ratio with the previous chart. There are twice as many bars for the same time period. The bandwidth routine parameters have not changed.

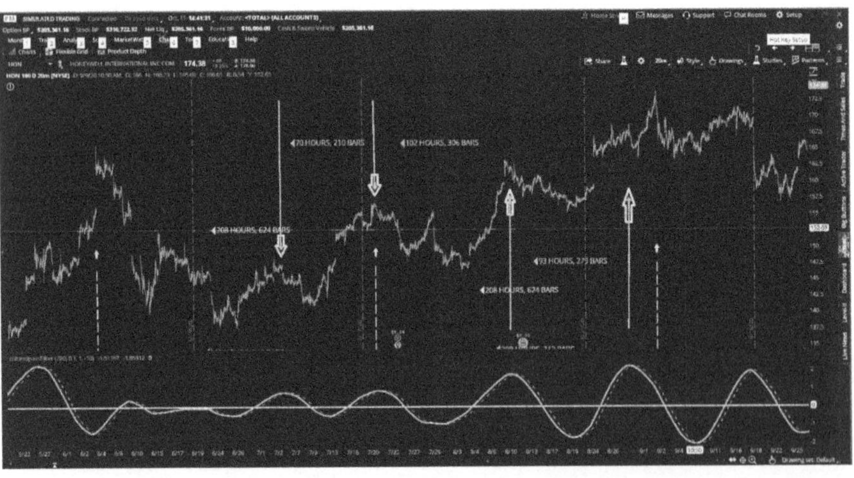

Figure 75: 20 Minute Chart (avg.: 88 Hours, 265 Bars)

The waveform generally lines up with major peaks that additionally appear more relevant at a lower scale. However, the distance between waveforms is somewhat uneven. You would expect the same number of bars for half the string distance at twice the frequency. The middle section was evaluated at 306 bars compared to 312 bars on the forty minute chart. However, the waveforms give us an average of 265 bars for this chart.

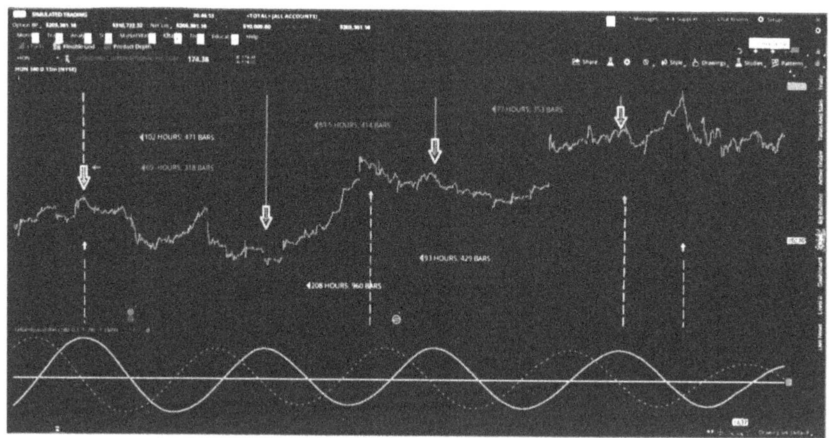

Figure 76: 13 Minute Chart (avg.: 78.5 Hours, 361 Bars)

In general, the 13 minute chart is in excellent agreement with the 20 minute chart. The scale is at 2 to 3 with the 20 minute chart and 1 to 3 with the 40 minute chart. The average number of bars is 361 and with the first third coming in at 318 bars compared to 312 bars for the 40 minute chart.

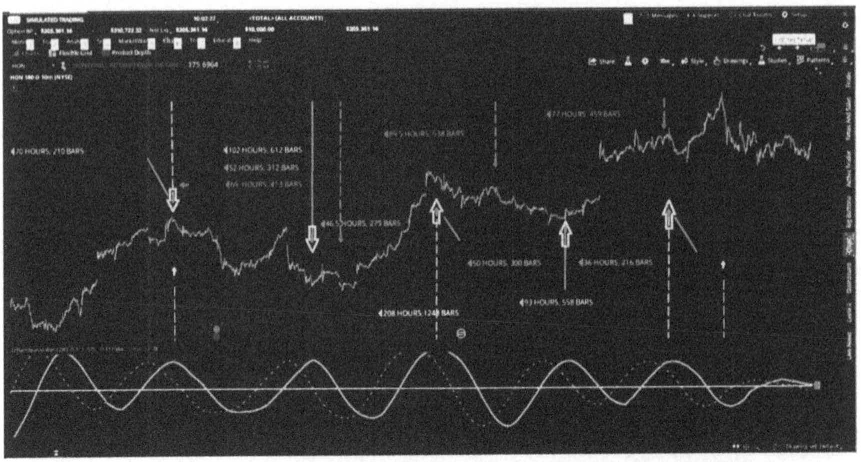

Figure 77: 10 Minute Chart (avg.: 46 Hours, 277 Bars)

This last chart ratio is 3 to 4 with the previous chart and 1 to 4 with the 40 minute chart. The average bar length is 277. Of the 4 sections shown, the first section has 312 bars exactly like the 40 minute chart. The second section has 279 bars which matches the last section of the 20 minute chart. The third section at 300 bars is almost the same as the third section shown of the 20 minute chart. The last section is short at 216 bars. If extended to the yellow line, it probably comes close to 312 bars.

Despite the filter routines seeming inconsistency in showing firm distances between events, the tops of the waveforms point to every up and down trade. Starting at the left yellow line and the peaks of the waveforms, the drop in the price is predicted by the second peak, the next rise in price is predicted by the third peak, the shorted fall is predicted by the fourth peak and the final peak stock price rises just short of the final high shown.

The 40 minute chart only indicated the peaks. With a 4 times expansion of the number of bars and a 4 times expansion of waveform frequency, every peak and valley was predicted by the waveform. The noise is gone. What remains is order; the order of the string lengths, the order of the musical overtones from the fundamental frequency, and the order of the number system – 1, 2, 3, and 4.

Annual discharges of the Nile at Aswan.

Period	Mean annual discharge (km³)	Standard deviation (km³)
1870–1899	110.0	17.1
1900–1959	84.5	13.5
1870–1959	92.6	19.8

Finally from the chapter on "Partial Order", let us return to Harold Hurst and the Nile River inflow project. The objective was to store water in times of drought for the Egyptian people, and also to build a dam to regulate the water flow into a new reservoir and decide how much water is needed to determine its size. Keep in mind that the Hurst exponent calculated was 0.72. Clearly, the annual cubic kilometer discharge was not a random event. Therefore, the normal distribution was not used to deduce the standard deviation.

$$(101) \qquad \sigma = \frac{\left(\frac{N}{2}\right)^{0.72}}{R} \qquad where$$

- $\sigma = standard\ deviation$
- $N = number\ of\ years$
- $R = range$
- $0.72 = Hurst\ exponent$

After 100 years of water records, the standard deviation was determined to be $18\ km^3$. To add to the table data, would Hurst's decision be helped by knowing the average or mean time in years where conditions of drought exist?

Based on the previous table, let us assume that poor water conditions exist for any water dispersions less than $84.5\ km^3$. The graph from Chapter 6 "Partial Order" shows the water dispersion levels for the Nile River over 137 years. Of those values, 49 exist under $84.5\ km^3$. Therefore, the mean time between poor water conditions is 2.8 years.

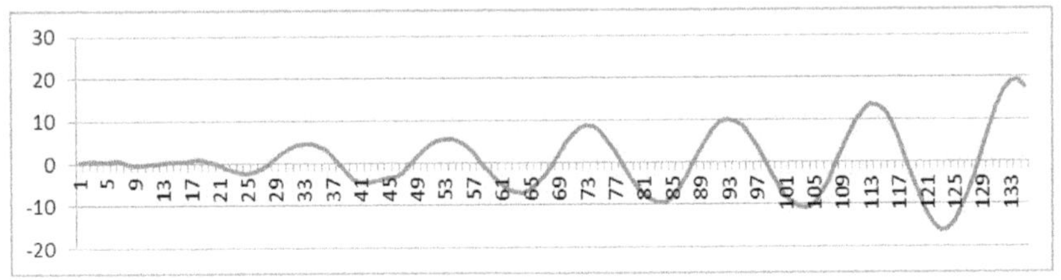

Figure 78: Nile River Mean Time for Poor Water Conditions

There is another way to calculate the mean value for poor water conditions. Despite the choppiness of the Nile River Chart, a bandpass filter can be applied for values between $45\ km^3$ and $84.5\ km^3$. The waveform above results in an angular frequency ω or a musical string length of 20 years.

$$(102)\qquad\qquad f = \frac{\omega}{2\pi}\qquad where$$

- $\omega = angular\ frequency$
- $f = time\ period\ or\ frequency$

The time period f for the mean value of poor water conditions is 3.18 years. With more data and a finer grain, the two methods – percentage of water results verses bandwidth – will become closer.

Chapter 13: Quantum Properties

Let us assume that a family takes a train trip of 150 miles from Davenport, Iowa to Chicago, Illinois. The train leaves the station at 12:00 noon and travels at 50 miles per hour. What time will the train arrive in Chicago?

In the world of Classical Mechanics, the answer is 3:00 PM. However, a quantum physicist would say that this answer is incorrect. The arrival time is a probability density of values. Perhaps the weather is bad. Perhaps a tree fell on the railroad track. Perhaps the train's speedometer was uncalibrated and the train got there early. There is a clash of reality between Classical and Quantum Mechanics. From the Heisenberg Uncertainty Principle:

$$(103) \qquad \Delta x * \Delta p \geq \frac{h}{2} \qquad where$$

- $\Delta x = change\ in\ position$
- $\Delta p = change\ in\ momentum$
- $h = Plank's\ constant = 6.62607004 \times 10^{-34}\ m^2\ kg\ /\ s$

If the train arrives at 3:00 PM, then the position x of the train is known and $\Delta x = 0$. This means that $\Delta p = \infty$. These two values cannot happen in Quantum Mechanics.

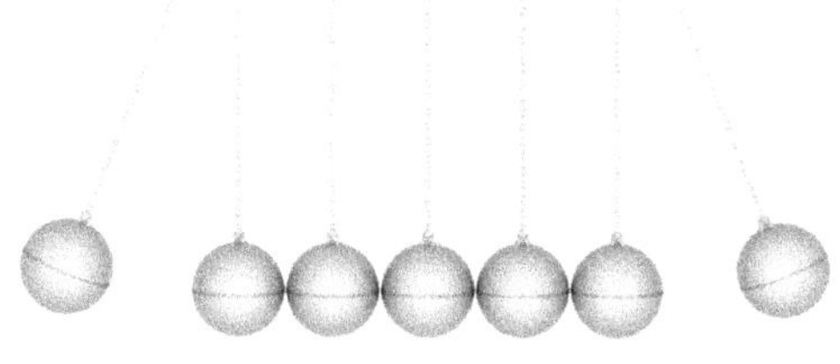

Figure 79: Newton's Cradle

In Classical Mechanics, knowing the position and momentum (velocity) of an object is key to solving a problem. Figure 79 is a diagram of Newton's cradle. It was named after the father of Classical Mechanics possibly because it requires the law of Conservation of Energy and the law of Conservation of Momentum to solve the problem of energy dispersion and height displacement.

$$(104) \qquad KE = kinetic\ energy = \frac{1}{2}mv^2$$

$$(105) \qquad PE = potential\ emergy = mgh$$

$$(106)\ \ E_T = total\ energy = KE + PE = \text{Conservation of Energy (constant)}$$

$$(107) \qquad p = momentum = mv$$

$$(108) \qquad \sum_1^n m_n v_n = 0 = Conservation\ of\ Momentum \qquad where$$

- $m = mass$
- $v = velocity$
- $g = gravity = 9.8\frac{m}{s^2}$
- $h = height$

As ball no. 1 strikes ball 2, in a perfectly elastic world the force will travel through balls 2, 3, 4, 5 and 6 and strike ball no. 7 that will knock the ball into the air. For example, let us assume that ball 1 has a mass m_1 that has twice the mass of balls 2, 3, 4, 5, 6 and 7. Also assume that ball 1 hit ball 2 with a velocity of 5 meters per second. How high h did ball no. 7 rise?

From Conservation of Momentum,

$$(109) \qquad m_1 v_1 = m_5 v_5 \qquad m_1 = 2m, m_5 = m$$

$$(110) \qquad v_5 = 2 * \frac{2m}{m} * v_1 = 4 * 5 = 20\ meters/sec$$

From Conversation of Energy,

$$(111) \qquad \frac{1}{2} m_1 v_1^2 = m_5 v_5 h$$

$$(112) \qquad h = \frac{1}{2} * \frac{2m}{m} * \frac{v_1^2}{v_5} = \frac{5^2}{20} = 1.25 \, meters$$

Contrast the laws of Classical Mechanics with the idiosyncrasies of Quantum Mechanics. While not complete, how would a mechanical engineer respond to the following properties?

- Not strictly derived equations from experimental observations, the scientific method. Equations are sometimes or partially mathematically derived.

- Momentum is a quantized value; not continuous.

- Position and velocity, the given values in Classical Mechanics, are not needed to solve Quantum Mechanics equations.

- All position values of quantum particles are standard deviations from a probability density distribution.

- All momentum values of quantum particles are standard deviations from a probability density distribution.

- The particle wave, the Schrodinger Equation, has quantized energy value solutions.

- Probability density values do not change with time. A wave solution and a time solution are independent of one another.

- In a vacuum, charge and magnetism are not zero ever.

This has been dubbed the Theory of Nothing.

- A hydrogen electron will stay in an orbital state forever unless excited by an outside influence.

- Because of the Heisenberg's Uncertainty Principle, position and momentum can never be zero.

- Two or more wave functions when added together and then respectively multiplied by constants or even other different functions can be solved with a single solution applied to all.

How does a mechanical engineer feel about this quantum soup? They should not feel bad. Quantum physicists don't understand it either. After 100 years of study, a typical quantum physicist will tell you that 95% of quantum physics is unknown. At the same time, they will tell you that quantum laws apply to Classical Mechanics also.

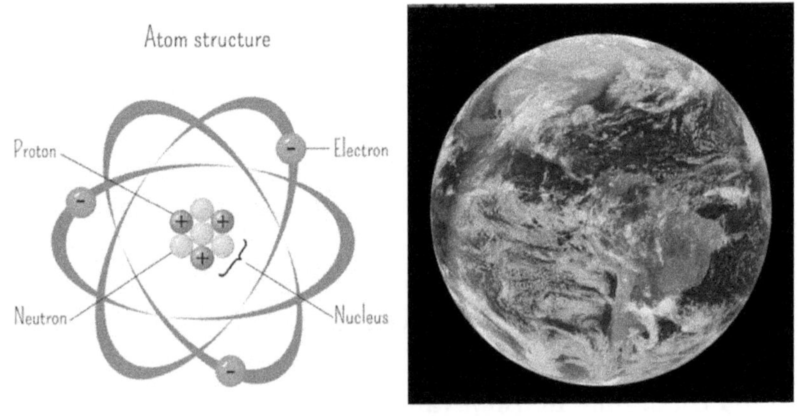

Figure 80: Electron Verses Earth Mass

For example, Quantum Mechanics says that all particles are

waves, and it also says that all masses are proportional to their wave lengths. Let us compare the wavelength of an electron with the wavelength of Earth.

Weighted Object	Mass	Wavelength
electron	$9.109\,383\,7015 \times 10^{-31}\ kg$	$1.23 \times 10^{-9}\ meters$
Earth	$5.9722 \times 10^{24}\ kg$	$8.06 \times 10^{45}\ meters$

The wavelength of the Earth is more than two times the distance of our solar system to the center of the Milky Way. Since an underground particle accelerator cannot measure anything finer than $10^{-23}\ cm$, the Uncertainty Principle tells us that the classical position of the Earth is assured as a localized measurement.

Now comes the hard part. Can Classical Mechanics and Quantum Mechanics even talk to each other? Let us review the Schrödinger wave equation.

$$(113) \qquad i\hbar\frac{\partial \varphi}{\partial t} = -\frac{\hbar^2}{2m}\frac{\partial^2 \varphi}{\partial x^2} + V(x)\Psi(x,t) \quad where$$

- $-\dfrac{\hbar^2}{2m} = KE = kinetic\ energy\ in\ the\ position\ domain$
- $V(x) = PE = potential\ energy\ as\ an\ undefined\ function$
- $i\hbar =$
  $a\ complex\ quantum\ momentum\ transition\ back\ to\ the\ time\ domain$

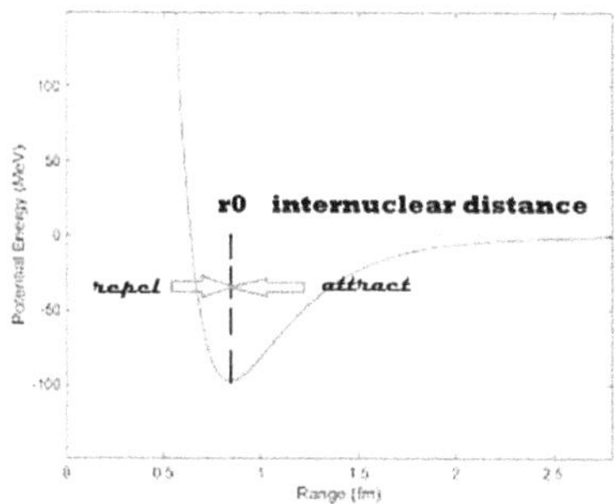

Figure 81: Electron Internuclear Distance

A solution, one of many, to the Schrödinger equation might look like this.

$$(114) \qquad \varphi(n, l, m)\alpha\left(\frac{Z}{a_0}\right)^{\frac{3}{2}} e^{-\frac{Zx}{a_0}} \qquad where$$

- n = *orbital number (quantized energy level)*
- l = *shape or path of orbital*
- m =
  magnetic quantum number (angular direction of orbital)

Based on this solution, the graph on the previous page demonstrates electron covalent bonding; that is, the electron attraction to and repulsion from the nucleus of an atom. The place where there is the highest magnitude (lowest y-axis value) is where the bond theoretically takes place. To the left of bonding is the area of repulsion and to the right is the area of attraction. Close to the bonding point representing the radial distance on the x-axis of the orbit, the orbit is beautifully circular. The orbit represents the Schrodinger waveform.

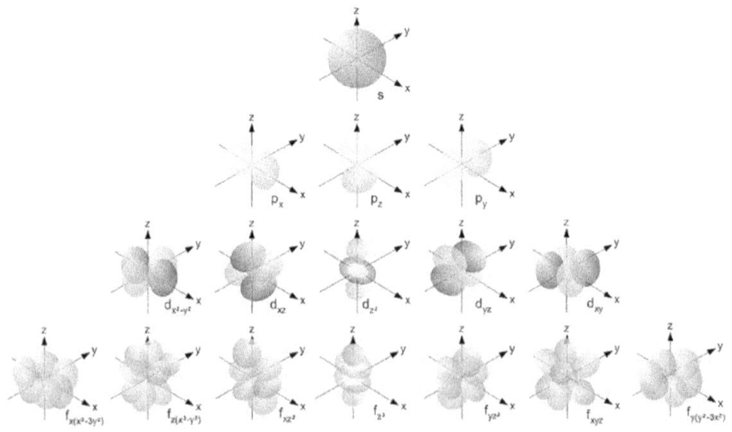

Figure 82: The Shape of Electron Orbits

In addition, chemical theorists have modeled the shape of the orbits. Any orbit only has two electrons with opposite spins. However, the n-level describes the energy orbit where the shape of the orbit exists. The orbits exist in three dimensional space. If there are 8 electrons at a particular energy level, then there are 4 orbits in different directions. These orbits have names: s, p, d, f, etc. The atomic number of a chemical specifies the number of electrons the element contains, and by analogy, the number of positively charged protons the nucleus contains. For example, look at the following descriptions:

$$(115) \qquad Cl(17) = chlorine = 1s^2 2s^2 2p^6 3s^2 3p^5$$

$$(116) \qquad Fe(26) = iron = 1s^2 2s^2 2p^6 3s^2 3p^6 4s^2 4d^6$$

The preface number describes the energy level n. The superscript describes the number of electrons at each level and at each shape. Adding across, the number of electrons will equal the element's atomic number. The shapes are from simple to more complex with s being the simplest shape.

Quantum physicists don't quite agree with this chemical model. Some believe that the orbit currently is undefinable. Others subscribe to the following pictures:

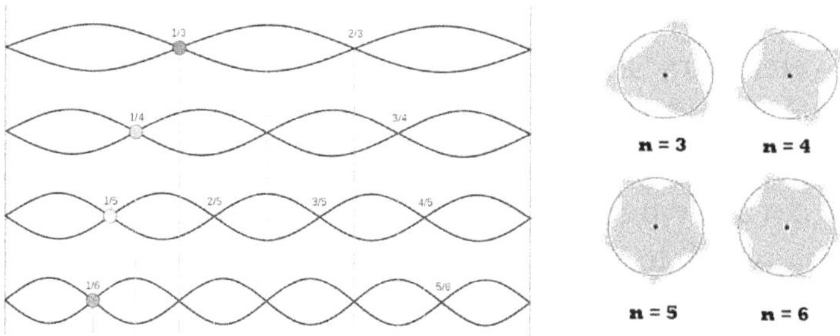

Figure 83: Quantum Electron Orbits

The electronic orbits around the nucleus must be complete waveforms. If the electron is in the 6th orbit (n=6), then there must be 6 complete waveforms. The orbit is not circular. It oscillates up and down.

Suppose we treat the orbit probability density waveforms classically. Recall that in Chapter 6 "Partial Order", the limited bandwidth of a cell phone was related to the musical series of overtones. Normally, the logic of overtones starts with the fundamental musical tone and telescopes upward with a series of overtones with faster frequencies. Just like the cell phone analysis where the fundamental musical tone was derived, is it possible to microscope downward to derive lower level frequencies? For example, would orbital cycle times of an electron at n=6,4,2 allow us to infer the cycle time for n=1? Further, would orbital cycle times for n=3,2,1 allow us to infer a smaller position density function at n=1/2? Could this be tested? Would the Uncertainty Principle have to be modified if the event were true?

	Potassium	Calcium	Scandium	Titanium	Vanadium	Chromium	Manganese	Iron		
	K	Ca	Sc	Ti	V	Cr	Mn	Fe		Mean
I. Level										
1	418.8	589.8	633.1	658.8	650.9	652.9	717.3	762.5		635.5125
2	3052	1145.4	1235	1309.8	1414	1590.6	1509	1561.9		1395.1
3	4420	4912.4	2388.6	2652.5	2830	2987	3248	2957		2843.85
4	5877	6491	7090.6	4174.6	4507	4743	4940	5290		4730.92
5	7975	8153	8843	9581	6298.7	6702	6990	7240		6807.675
6	9590	10496	10679	11533	12363	8744.9	9220	9560		9174.967

Above is a table of ionization energy levels for the fourth row of elements in the periodic table and with atomic numbers from 19 to 26. An ionization energy level describes the energy in picojoules $(10^{-12}\ joules)$ required to break an electron covalent bond. The table has 6 levels. Level 1 is the energy required to free one electron. Level 6 is the energy required to free 6 electrons. Electrons are removed from the outer shell or outer orbit only. A stable element is one that completely fills the outer orbit. The lowest orbit only contains

2 electrons. Each succeeding orbit can contain 8 electrons. Elements with the atomic numbers, 2, 10, 18, and 26 like iron Fe are stable or inert and do not require a partner. All other elements have free electrons and are available to form molecule strings.

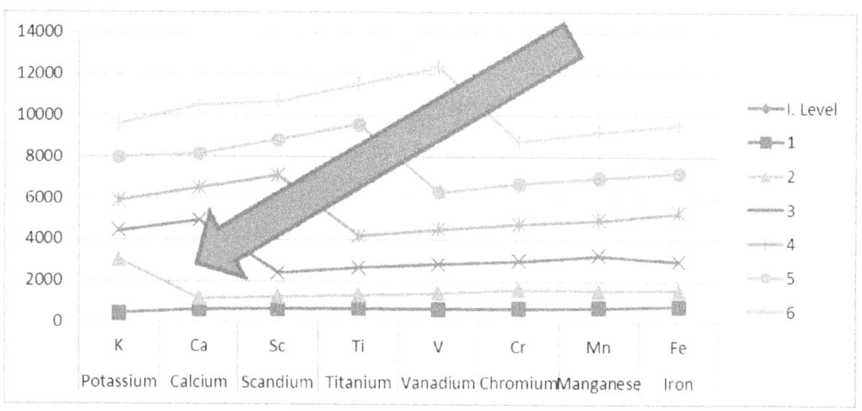

Figure 84: Ionization Energy Levels for Electron Separation

The above chart plots the energy levels for all the elements. Note the fall of the levels as indicated by the arrow. Each element has a different number of electrons in the outer shell. As the outer shell is stripped of electrons, the energy level falls to the next lower level. Also, as the number of electrons available increases from left to right, the number of protons also increases leading to a stronger nuclear force. With only 6 levels shown, chromium (Cr), manganese (Mn), and iron (Fe) have yet to experience an outer shell collapse.

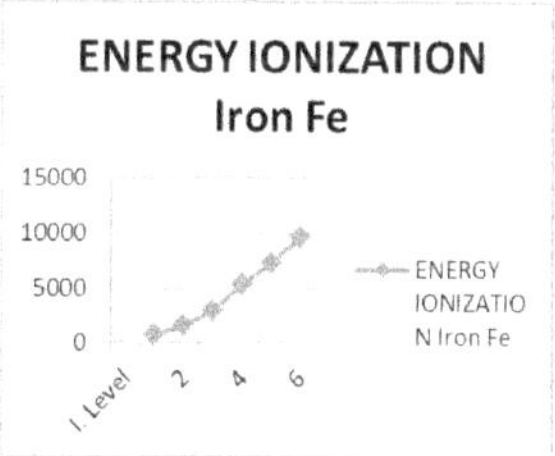

Figure 85: Energy Ionization of Iron (Fe)

In general, the ionization energy graphs for each element (iron on the right side) look like 45 degree straight lines. This suggests proportional energy values between each level. The following table shows the mean or average values for each level and the estimated ratios between each level. Two mean values are listed. The first value is for all the data at each level. The second list of mean values was averaged only after the energy level drops due to the stripping of electrons in the outer shell. Level 6 only has 3 values while level 1 uses all its values.

Energy Level	Ionization Mean	Ionization Mean After Energy Drop	Approximate Ratio Between Levels
1	635.5	635.5	
			5:2, 2:1
2	1602.2	1395.1	
			2:1
3	3299	2844	
			3:2
4	5389	4731	
			4:3
5	7723	6808	

While the ratios are approximate, they begin to look like the overtone ratios of the musical scale. Keeping in mind that the ionization energy measurements listed are from a probability density distribution, and assuming a random distribution with random energy measurements, let us look at the energy levels for iron Fe. Starting with level 6, the objective is to use the approximate ratios to predict the ionization energy values for levels 1 through 5. In effect, the procedure microscopes down.

Energy Levels for Iron (Fe)	Assumed Ratio Applied	Predicted Energy Level	Actual Energy Level	Percent Error
6	4:3		9560	
5	4:3	7170	7240	1.0 %
4	3:2	5378	5290	1.7 %
3	2:1	3585	2957	21.2 %
2	5:2, 2:1	1792.5	1561.9	14.8 %
1		717, 896	762.5	6.3%, 13.5%

While the predicted values for levels 4 and 5 may be well within the probability density distributions of the ionization energy values, at least in this case the results were not good. The objective was to extrapolate a quantum value that was difficult to obtain in the laboratory and see if the value helps the predictive math. Is it reasonable to separate the quantum world and its randomness from the classical world and its discrete whole-number proportions? Is it reasonable to assume that all aspects of chaos theory and the proportions therein do not apply to quantum theory?

Chapter 14: Ancient Geometry

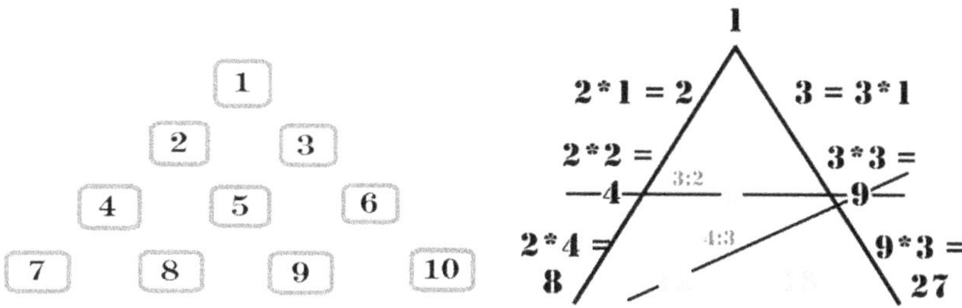

Figure 86: Pythagorean Tetractys   Figure 87: Plato's Lambda

One plus two plus three plus four equals ten. Pythagoras arranged the 10 numbers in the form of a triangle with four rows. This famous arrangement is called the Pythagorean Tetractys.   On the right is an arrangement called Plato's Lambda. Multiplication by 2 happens down the left side and multiplication by 3 happens down the right side. By drawing lines through the numbers, the discovered ratios are 2 to 1, 3 to 1, 3 to 2, and 4 to 3. These are the major musical ratios of the octave, the dominant major 5th, and the subdominant major 4th.

Apparently, Pythagoras and Plato travelled their world including Egypt, and then back in Greece, their teachings revealed hidden secrets. "The Temples of God" kept numbers and geometric shapes from their population as hidden knowledge. Pythagoras and Plato were among the first mathematicians to reveal knowledge to their select students about the secrets that they had learned.

Figure 88: Hieroglyphic Template

The image above is a template for all ancient knowledge. It is a little dramatic in that there should only be 64 squares. The square, the triangle, and the circle were the shapes used on clay tablets to form the cuneiform languages of the Egyptians, Sumerians, Assyrians, Babylonians and other ancient civilizations. Imagine a stick being sharpened to poke holes in a laptop clay tablet. Let us say that we want to count to 50. There simply is not enough room to poke 50 holes. This was their solution.

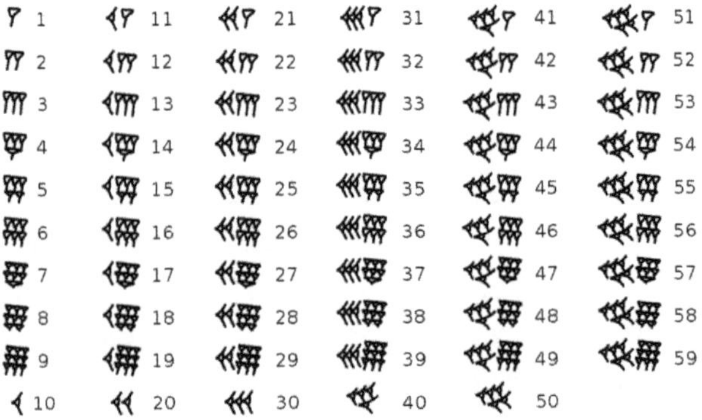

Figure 89: Babylonian Numbers

The hole mark is triangle shaped and the line drawn on the template is to provide more room for counting. When they reached the number 10, they turned their stick sideways to create a triangle arrow. How about their cuneiform language?

Figure 90: Hieroglyphic Words

In general, the cuneiform languages conform to the rather rectangular images formed from the template. The languages allowed for written communication and trading between cities. The measurements used led to a sexegesimal system.

There actually are three digits in the example. From right to left:

- *Digit* 1 = 60^0 = 1
- *Digit* 2 = 60^1 = 60
- *Digit* 3 = 60^2 = 3,600

	3,600	X	2	=	7,200
	60	X	22	=	1,320
	1	X	43	=	43
					8,563

Figure 91: Sumerian Addition

In each case, the maximum number that can exist in each digit is 59. Above 59, the quantity overflows into the next digit.

If a society wanted to measure something or divide a parcel of land among a group of farmers, they would tend to use anatomical measurements like a foot or a cubit. A cubit is a unit of measurement from the elbow to the tips of the fingers. A man's stride, defined as stepping left-right, produces a double cubit, or approximately a yard. 200 cubits equals 100 yards. The farming parcels were divided equally with a common length and width, say 200 cubits. The units would be whole numbers. A third of an acre was a foreign concept.

The arithmetic of the ancients centered around whole numbers and whole number fractions:

$$(117) \qquad Number\ Set = \left[\ldots \frac{1}{8}, \frac{1}{7}, \frac{1}{6}, \frac{1}{5}, \frac{1}{4}, \frac{1}{3}, \frac{1}{2}, 1, 2, 3, 4, 5, 6, 7, 8, \ldots \right]$$

Notice that the arithmetic numbers centered around the "God" number 1 (unity). There was no zero; there were no negative numbers, there was no infinity. Counting meant representing objects that existed. There were no zero apples. There were no negative crops.

Without division by zero, there would be no infinity, and by analogy, there would be no mathematical paradoxes or singularities. Let us solve the quadratic equation using yesterday's rules.

$$(118) \qquad ax^2 + bx + c = 0$$

$$(119) \qquad x = \frac{-b \pm \sqrt{b^2 - 4ac}}{2a} \quad and \quad a, b, c > 0$$

$$(120) \qquad -b \pm \sqrt{b^2 - 4ac} > 0$$

No matter how you slice it, $\sqrt{b^2 - 4ac} < b$. Therefore Equation 4 is negative and the quadratic equation cannot be solved by ancient arithmetic.

Figure 92: Squaring the Circle   Figure 93: Archimedes Pi

Prior to calculus, there was no direct way to calculate the 100 digits of pi (π). In fact, the squaring of the circle became the most worked on problem in early mathematics history. The objective was to calculate the area of a circle, a curved object. However, Isaac Newton had the advantage of knowing that a half a circle circumference equals pi. By using the formula for the arc length of a semicircle, which was derived using Pythagoras's Theorem:

$$(121) \qquad \int_{-1}^{1} 1/\sqrt{1 - x^2} \; \partial x = arc \sin(x) = \pi$$

The difficulty in accepting this answer is that you already know the answer before you start and used it in your definite integral formulation. The real hero was Archimedes of Syracuse (279 BC to 212 BC), a Greek mathematician. Archimedes drew a square both inside a circle and outside a circle to try to estimate both the area and circumference of a circle. By assigning unit values to the sides, he knew that the circumference was less than four. However, there were gaps between the square and the circle. To narrow the difference down, he tried 5 sides, then 6 and finally 96 sides by manual calculation. He arrived at the boundary limits as shown and a value for pi = 3.14.

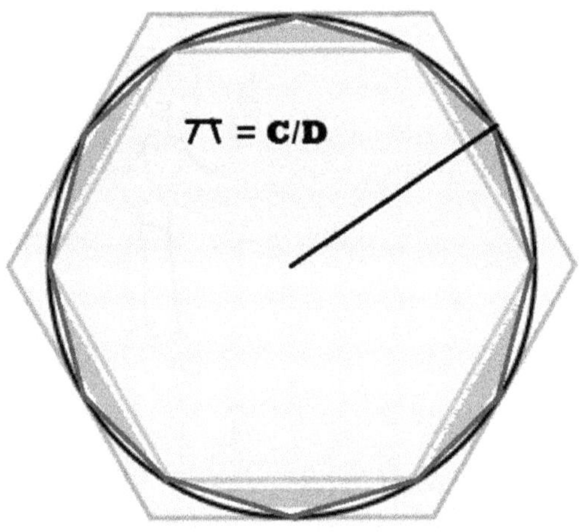

Figure 94: Euclid's Approximation of Pi

In our time, there are a number of Euclidean methods to approximate Archimedes method. Euclid of Alexandria (325 BC to 265 BC) wrote his treatise on geometry before Archimedes discovered pi. Above is an octagon drawn inside the circle. Given the formula for pi = cirumference/diameter indicated above, six sides with unit distance inside the circle means that pi > 3. The 12-sided dodecahedron shown allows both a diameter and a circumference to be calculated based on the formulas shown. Looking closely, there are two 90 degree triangles that allowed distances to be calculated based on the Pythagorean Theorem.

n	S	S/2	a	b	New S	C	C/D
6	1	0.5	0.86603	0.13397	0.51764	6	3
12	0.51764	0.25882	0.96593	0.03407	0.26105	6.21166	3.10583
24	0.26105	0.13053	0.99144	0.00856	0.13081	6.26526	3.13263
48	0.13081	0.0654	0.99786	0.00214	0.06544	6.2787	3.13935
96	0.06544	0.03272	0.99946	0.00054	0.03272	6.28206	3.14103
192	0.03272	0.01636	0.99987	0.00013	0.01636	6.2829	3.14145
384	0.01636	0.00818	0.99997	3.3E-05	0.00818	6.28312	3.14156
768	0.00818	0.00409	0.99999	8.4E-06	0.00409	6.28317	3.14158
1536	0.00409	0.00205	1	2.1E-06	0.00205	6.28318	3.14159

The parameters shown in the diagram were calculated in the table. For n=96, pi=3.141 with three places of accuracy. For n =1536, pi=3.14159 with five places of accuracy.

The "Temples of God" hid basic arithmetic and geometry from their constituents. In modern times, this spawned the phrase "Sacred Geometry". Generally, the secrets are really the basic arithmetic and geometry. Sacred geometry centers around the arithmetic mean, harmonic mean, and geometric mean.   The geometric mean is a way to solve non-linear problems. The growth patterns flow through $\sqrt{2}$, $\sqrt{3}$, and $\sqrt{5}$.

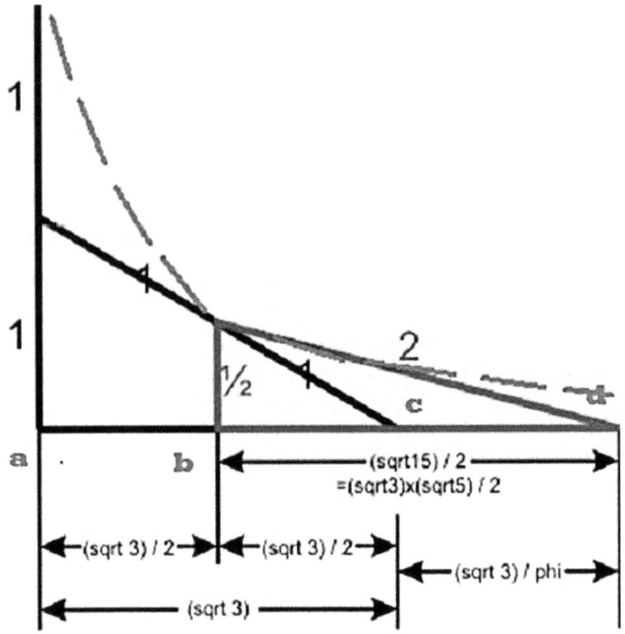

Figure 95: Knife Triangles

An interesting geometry called the "knife arrangement" by its author Michael Schneider, *The Golden Ratio at the Dinner Table*, bisects each line with length 2 at the midpoint and lays the "knife" end on the X-axis. Given Pythagorean triangles, there are relationships with $\sqrt{2}$, $\sqrt{3}$, $\sqrt{5}$, and at the very end phi($\varnothing$), the Golden Ratio. In addition with a few more knives, an exponential function of the form e^{-kt} can be traced from the top of the Y-axis through the midpoint of the knives. Let us consider the first triangle (black) and the points on the x-axis which we will label a,b,and c.

$$(122)\qquad \frac{a-b}{b-c} = \frac{\frac{\sqrt{3}}{2}}{\frac{\sqrt{3}}{2}} = 1 = arithmetic\ mean = \text{average growth rate}$$

For the second triangle (red), let us add the end point d.

(123)

$$\frac{a-c}{b-d} = \frac{\frac{\sqrt{3}}{\sqrt{3}}}{\varnothing} = \varnothing = geometric\ mean = geometric\ growth\ rate$$

Finally, the parallel growth of the second triangle is examined.

$$(124) \qquad \frac{1}{b-c} + \frac{1}{c-d} = \frac{\frac{1}{\sqrt{3}}}{2} + \frac{\frac{1}{\sqrt{3}}}{\emptyset} = \frac{\emptyset + 2}{\sqrt{3}} = 2.09 = harmonic\ mean =$$
$$growth\ acceleration$$

The fundamental principle of these means, used proportionally in ancient times, is balance in motion, a concept that is at the heart of the Constitution of the United States.

Chapter 15: Musings on Goodness

There has to become a point in this book where an attempt is made to connect the dots. Back in ancient times, knowledge was hidden all in the name of control. Cuneiform numbers were hidden. Simple geometric forms were hidden. After the bible was formed around 400 A.D., it was removed from society for hundreds of years. Scientists like Galileo, Newton, and Darwin were reluctant to release their writings. Even politicians today have hidden agendas.

In order to connect the dots, one must put down any thought of scholarly robes and become a sage, a mystic, a religious person, a societal journalist; in other words, the exact opposite of a scientific person. These musings on goodness are an attempt to put a positive spin on our surroundings.

One of the first orders of civilization for a rational being is to look within yourself. If the world is random and your destiny is by chance, how can you do this? Proportion is how we look at things. Balance in motion is an objective. If you are standing still, what can you accomplish? When a monk meditates, he is looking at a perfect reflection of self. In electrical engineering, this is called impedance matching. For a sacred person, the reflection of God is called unity.

$$(125) \quad 1/(1+1)/(1+1)/(1+1)/(1+1)/\ldots = \emptyset = 1.618 = \textit{The Golden Ratio}$$

Your divine nature is in matching the Golden Ratio. The Fibonacci Spiral is an example of Golden Ratio proportions, is found everywhere in nature, and represents balance in motion.

Proportions can be tricky. There are three types of proportions:

$$(126) \qquad Formulative = \frac{a}{b} = \frac{c}{d}$$

$$(127) \qquad Relative = \frac{a}{b} = \frac{b}{c}$$

$$(128) \qquad Dimensional = \frac{a}{b} = \frac{b}{a+b}$$

Now let us try three examples to determine which one is true:

- Example 1: $\dfrac{apples}{oranges} = \dfrac{orchard}{grove}$
- Example 2: $\dfrac{apples}{oranges} = \dfrac{oranges}{orchard}$
- Example 3: $\dfrac{apples}{oranges} = \dfrac{oranges}{apples+oranges}$

The first example implies that apples are grown in an orange grove. Frankly, this is a theory of relativity problem as well as a political science problem. The seeming equality causes a widening of dimension that may not reflect self or the real truth. While apples do grow on trees, a partial derivative of each side represents chaotic motion; hence, the particle motion of $\sqrt{2}$. Too often, this type of proportion represents the scientific principle; that is, the dispassionate from a distance look at the evidence.

The second example is not dispassionate; it has a conscience in the middle. Are apples related to an orchard? How do you know? According to Webster, the root of conscience is "to know". A conscience is a living entity with a memory. Since there is a "memory" in the middle, the proportion is not dispassionate and does not represent the scientific method. How ironic, a proportion that portends "to know" or "the truth" is not scientific. Since most formulas in science model "dead" systems -- that is, mechanical knowledge – this proportion is about an "open" system derived through the eyes of life. The observations are compared with a knowledge base (memory), and adjustments are made in personal knowledge to arrive closer to self-awareness. Since there are three variables, $\sqrt{3}$ is the derivative of the three variables. It is an involution of knowledge.

The third example is a perfect reflection of how apples and oranges are related. They are both fruits. They came from "woody" plants; hence, they can belong to the same category of trees. Whether or not you realize it, you have arrived at a higher dimension. In addition to the differential information, apples and oranges have revealed integrated facts. Apples plus oranges is like a Venn diagram integrating their respective properties as fruit. Because there is both a differentiation and an integration, the proportions are influenced by $\sqrt{2}$ and $\sqrt{3}$. These sizes with only two variables are related to $\sqrt{5}$. We are getting closer to the Golden Ratio.

Let us relate these proportions to the transfer of knowledge. When you meditate, in essence, you are relating what you know to yourself or to your fundamental being. You are

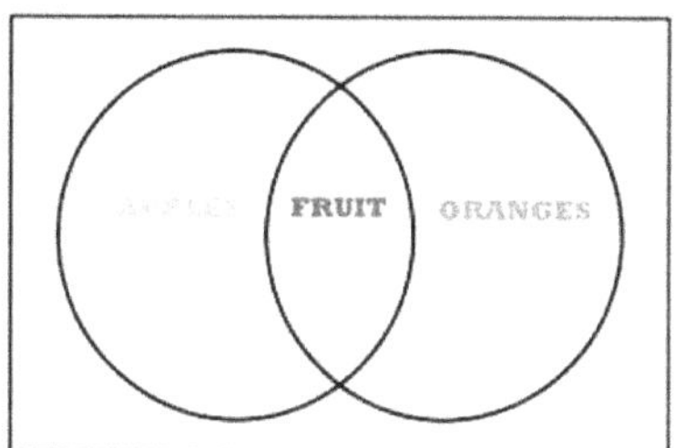

Figure 96: Dimensional Proportion Venn Diagram

taking time-based knowledge, facts in motion, and you are relating them to a fundamental standard – you! It is like a Fourier Transform, where time-based data are converted to frequency-based data. You are taking waves or wavelets and converting them to circles. Circles repeat. There is no time. In our example, apples and oranges are wavelets; the fruit Venn diagram is the encompassing circle.

There are many paradoxes in science because there are too few dimensions. Which came first, the chicken or the egg? Whatever side you choose, your deductions lead to an infinite loop. The solution involves mathematics with zero and infinity. Draw a Venn diagram. In essence, the chicken and the egg meet in the middle. You have converted a one dimensional loop into a two dimensional problem. In the middle, the chicken and the egg are members of the same category. This is how infinite loops are broken. You must go up one dimension. In a two dimensional space, if there are more chickens than eggs, then there is a vector that tilts toward more chickens.

Global warming is a one dimensional problem, or so the pundits would have us believe. If the high in St. Louis reaches 100 degrees, it must be global warming despite the fact that it reaches 100 degrees almost every summer. A core sample taken from the frozen ground in Greenland shows an increase in bacteria. Yet bacteria spur plant growth that in turn becomes a receptacle for carbon dioxide and a chloroform factory for new oxygen and thus cools the earth. Think of the top 10 out of 100 parameters that might affect global warming. If these factors cannot be correlated, then having a change in a core sample changes the direction and magnitude of a 10-dimensional vector, but it does not directly lead to a one dimensional conclusion about global warming.

There are universal laws that help define order. In terms of energy, the conservation of energy is a universal law. Total energy is a constant, but potential verses kinetic energy can shift. The Second Law of Thermodynamics is based on a closed system with no inputs. If you mix two glasses of water at different temperatures, say boiling verses freezing, the total will be an average of the two temperatures; i. e., 212 degrees Fahrenheit and 32 degrees Fahrenheit will average to 122 degrees. Kinetic energy from the boiling water will thaw the freezing water to produce two cups of water with a potential energy of 122 degrees, exactly the same energy as the total of the two cups separately. The second laws says that out in deep space the temperature of the water will reach absolute zero; -273 degrees Fahrenheit. (Actually, deep space has a temperature of plus 2 degrees Kelvin; therefore, the water will never go below that temperature.)

As the temperature of the water drops, the molecules slow

down and separate. At near absolute zero, the molecules are separated by a space the size of two cups; a very empty space in the micro world. The water becomes disorganized. The Second Law of Thermodynamics is about becoming disorganized, about cooling and about the loss of potential energy. In other words, it is about death; the death of the process.

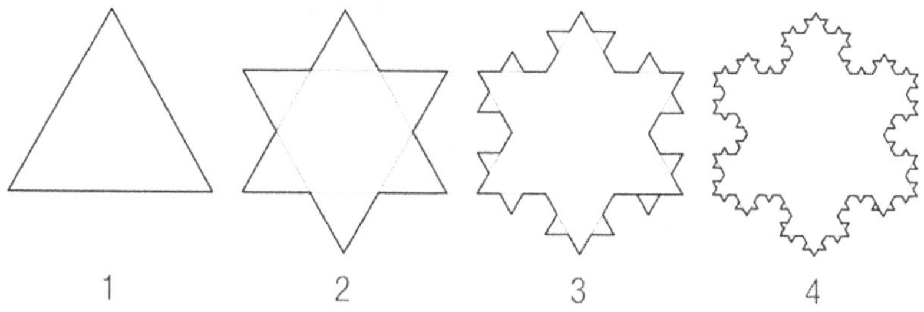

Figure 97: Koch Snowflake Potential Energy Growth

For a two dimensional problem, potential energy PE is the area under the curve and kinetic energy KE is a difference or differential energy in the direction of the vector energy. Kinetic energy is thought of as a one-dimensional entity. Let us take an abstract problem like the Koch Snowflake and pretend that the area of the snowflake is potential energy and the algorithm or process is the organizing kinetic energy.

Given an arbitrary length of an equilateral triangle, the side is broken into three parts, The snowflake starts as an equilateral triangle that is inserted in the middle third of the line with the length of each side s.

$$(129) \qquad area\,triangle = A = \frac{1}{2}bh = \frac{1}{2}s\sqrt{s^2 - \left(\frac{1}{2}s\right)^2} = \frac{\sqrt{3}}{4}s^2 = PE$$

For the second interation n=2, where every line including the first triangle is broken into three equal parts and a new equilateral triangle inserted, then in iterative fashion $s_2 = \frac{1}{3}s_1$ for 3 new sides.

$$(130) \quad A_2 = \frac{\sqrt{3}}{4}\,s^2 + 3\,\frac{\sqrt{3}}{4}\left(\frac{1}{3}s\right)^2 = \frac{\sqrt{3}}{4}\,s^2\left(1 + \frac{3}{9}\right) = PE$$

For $n = 3$,

$$(131)\quad A_3 = \frac{\sqrt{3}}{4}\,s^2\left(1 + \frac{3}{9}\right) + 3 * 3 * \frac{\sqrt{3}}{4}\left(\frac{s}{9}\right)^2 = \frac{\sqrt{3}}{4}s^2\left(1 + \frac{3}{9} + \frac{3*4}{9^2}\right) = PE$$

For the nth term,

$$(132) \quad A_n = \frac{\sqrt{3}}{4}\,s^2 \sum_{1}^{n}\left(\frac{3*4^{n-1}}{9^n}\right) = PE$$

The growth in kinetic energy available for just one side of the initial triangle is equal to Equation 132. Kinetic energy and heat are related. Energy is the ability to do work in joules (newton-meter) or foot-pounds. The English units have an advantage, as the ancients knew, in that there are no round-off errors with whole number fractions. The inference according to Plato in the Timeas is that the fall of fictional Atlantis was due to round off error using powers of ten or the decimal system. Calories measure the ability to heat water to a new temperature. Calories are a unit of energy. One foot-pound equals 0.3240 small calories. The calories on food are large calories or 1000 small calories per large calorie.

While this is a hypothetical example, let us assume somewhere in the atmosphere that a snowflake was forming by absorbing the surrounding humidity and forming crystals at a freezing temperature. The conservation of energy is preserved in that the energy lost in the atmosphere is added to the snowflake. Therefore for Equation 130 and n=1, the increase in energy starting with an equilateral triangle with sides one foot in length is 0.0481 ft-lbs or 0.0156 calories. Calories are heat. The Koch snowflake as it grew became warmer. This seemingly violates the Second Law of Thermodynamics. Why?

First, the Koch snowflake is not a closed adiabatic system. It is an open system with energy from an outside source; in this case, pencil energy. Therefore, the Second Law is not violated.

However, does it not stand to reason that a process that increases the energy and temperature of a system or resource is a living system or resource? Should there not be a scientific principle that models this growth? In the Old Testament, God says "I am that I am." I exist so that I might recreate myself. My heavenly job is creation. Should not creation have a scientific model or does mankind have to invent metaphysics to speak of creation?

It is hard to imagine Newton or Gauss or Maxwell or Einstein answering no to these questions. Calculus is the mathematical basis of energy. When you differentiate you are measuring a kinetic source. When you integrate you are measuring a potential source.

The First Law of Thermodynamics says that energy E must be preserved.

$$(133) \qquad \Delta E = T\Delta S - P\Delta V \qquad where$$

- $T = temperature$
- $S = entropy$
- $P = pressure$
- $V = volume$

The snowflake has been simplified in that this is a two-dimensional exercise.

Therefore, $\Delta V = 0$. For $n = 1$,

$$(134) \qquad \Delta E = \frac{\sqrt{3}}{4} s^2 = T\Delta S \qquad \text{where}$$

$$(135) \qquad \Delta T = \frac{calories}{calories\ per\ degree\ Celcius} = \frac{0.0156}{1} = 0.0156\ degrees\ Celcius$$

The increase in energy means that entropy is decreasing. Given side $s = 1/3$,

$$(136) \qquad \Delta S = -\frac{\Delta E}{\Delta T} = -\frac{\frac{\sqrt{3}}{4}s^2}{\Delta T} = -\frac{0.0481}{0.0156+273.15} = -0.000176\ joules/Kelvin$$

This negative change in entropy implies more organization, a better snowflake. However, the Second Law of Thermodynamics makes it clear: A change in entropy can never decrease. The Second Law gets around this in two ways; by 1) equating the atmosphere with a decrease in entropy, and 2) by calculating a new maximum entropy that allows the current relative change in entropy to be zero.

From the Second Law Corollary discussed in Chapter 11, the intropy P value change would be:

$$(137)$$
$$\Delta P = k\Delta T = 1.38x10^{-23}(0.0156 + 273.15) = 3.77x10^{-21} joules$$

The increase in potential energy signals a life spiral and not a death spiral. The randomness of entropy flies in the face of a life process that is organizing.

For the holistic thinkers of ancient times, John 1:1 defines goodness as well as anyone: "In the beginning there was the Word, and the Word was with God, and the Word was God."

$$(138) \qquad Word = 1 + \frac{1}{Word} = \emptyset = The\ Golden\ Ratio$$

Ancient geometry describes a creation process of involution (integration) and evolution (differentiation) involving $\sqrt{2}, \sqrt{3}, \sqrt{5},$ and $\varnothing$.

$$(139) \qquad \varnothing = \frac{1+\sqrt{5}}{2}$$

A Fibonacci series is a relative process of ancient proportions that results in a series of values related by the Golden Ratio. If we have a process, say good nutrition, and we stick to the process, we can take advantage of these ancient ratios. A bounded random process leads to the Sierpinski Gasket. The randomness does not lead to the Second Law of Thermodynamics because there is an organization process happening. Order is being created. Therefore, if a steady process is followed and it is harmonious with its surroundings, the result is growth using the proportions of growth. Good nutrition leads to good health. The less nutrition you practice, the less health you will have because the process has become unbounded.

Growth or decay is defined actually by two Spirals, kinetic verses potential energy. The spirals are additive. Given growth from birth to fullness, the small additions amount to a well behaved exponential curve at birth. However, death requires large additions initially. This produces a wiggling but flat top followed by a sharp drop off as the death curve becomes bigger.

Figure 98: Kinetic and Potential Energy

This imbalance is responsible for a top that defines the limit of the growth cycle and a possible evolution or gap due to the irregularity of the death cycle.

Figure 99: Kepler Telescope Journey to Find Life

Between 2009 and 2018, the Kepler Space Telescope traveled outside our solar system and discovered 2,600 planets in the Milky Way ranging over 500,000 stars. Of the planets, around 5 percent were approximately the size of the Earth. Using the Earth as a barometer, the possibility of life outside our solar system, even intelligent life as we understand intelligence, exists.

Johannes Kepler (1571-1630), German astronomer, discovered three laws of planetary motion. The Third Law states that the period of an orbit squared L^2 is proportional to the radius cubed R^3 of the orbit (the semi-major axis of the ellipse). Consider further that the speed the planet travels (relative to the sun) is the kinetic energy of the planet, and that the area enclosed by the elliptical orbit is a definition of the potential energy of the planet (caused by the gravity pull and the energy of the sun). Adding Newton's law of gravity to the proportion yields:

$$(140) \qquad L^2 = \left(\frac{4\pi^2}{G(m_1 + m_2)} \right) R^3$$

For the distance between the Earth and the sun,

$$(141) \qquad R^3 = L^2 \left(\frac{G(m_e + m_s)}{4\pi^2} \right) \qquad where$$

- $L = orbit\,period = 1\,year = 31,557,600\,seconds$
- $G = gravity = 6.674x10^{-11}\,\frac{m^3}{(kg \cdot s)}$
- $m_e = mass\,of\,Earth = 5.9736x10^{24}\,kg$
- $m_s = mass\,of\,sun = 1.989x10^{30}\,kg$
  - (1) $R^3 = 3.36236x10^{21} * 995.844x10^{12} = 3,348,386x10^{30}$
  - (2) $R = 149,604,677\,km = 92,960,038\,miles$

There is hardly anything more majestic than the Music of the Spheres. These orbiting planets are laying out the musical language of Nature; the overtone proportions of cyclic behavior. Kepler's Third Law of motion is the three-halves $\left(\frac{3}{2}\right)$ law. For the well-tempered scale, the three-halves law is the circle of fifths. Over seven octaves, the 12 keys of the chromatic scale of notes circle around to home base.

$$(142) \qquad 2^7 \cong \left(\frac{3}{2}\right)^{12}$$

Figure 100: Fractal Nature

In the mathematics of fractal geometry and chaos theory, a fractal is a self-similar subset of Euclidean space whose fractal dimension strictly is more precise than its topological dimension. Fractals appear the same at different levels, as illustrated in successive magnifications of the Mandelbrot set. For the first time, chaos theory presents a way to describe living objects and dynamic processes.

Fractals are a bridge to a higher dimension. A fractal describes the relation between a branch and a leaf, a canyon and a river, your elbow and your arm, a price chart and time. Nature has far more intelligence and memory than is commonly understood. Whether or not people survive a tornado, Nature knows how to rejuvenate itself.

If an Einstein particle out in deep space practices random motion and does not run into anything, then it will do no work. It will remain in an open unbounded world in chaotic motion forever. However, as soon as it bumps into another particle, it will be bounded by the attractive forces of the particle. A process, a particle dance, will spring to life. Eventually, the random process of both particles will order themselves. If there is an overarching argument against the Uncertainty Principle, this is it.

Chapter 16: The Nature of Reality

Figure 101: Aswan High Dam

Late in life, Harold Hurst argued vehemently that the Aswan Dam should be built higher than a Normal Distribution or randomness could predict. Not only did his exponent densities predict this, but the amount of rainfall was a significant factor. Hurst discovered that while rain and flooding might average normally over long periods of time, it was the sequence of weather that proved his skeptics wrong. In his book <u>The (Mis) Behaviour of Markets</u> published in 2004, Benoit Mandelbrot commented on Hurst's correlations:

"How much does the past shape the future? A moral philosopher would phrase it this way: Is it fate that determines our course, or do we choose our paths afresh with each new decision? A mathematician trades in another terminology: Is one event dependent on another, or independent from it? If Event B is dependent on Event A, then A's occurrence changes the odds of B happening. If a basketball player sinks two shots in a row, evidence suggests, odds are greater that his third shot will also score. By prowess or psychology, a player can have "hot" streaks, successive shots are, to some degree, dependent on another. But how long will his scoring streak last? Is it broken after just one miss? Two? Five? Over how many shots, precisely, does the "hot-hands" effect linger? ... As any chartist has learned to his sorrow, the most random and independent events can spontaneously appear to form patterns and cycles."

In his book <u>Science and Human Transformation</u>, published in 1997, Professor William Tiller, Head of the Material Science Department at Stanford University, writes about human consciousness C:

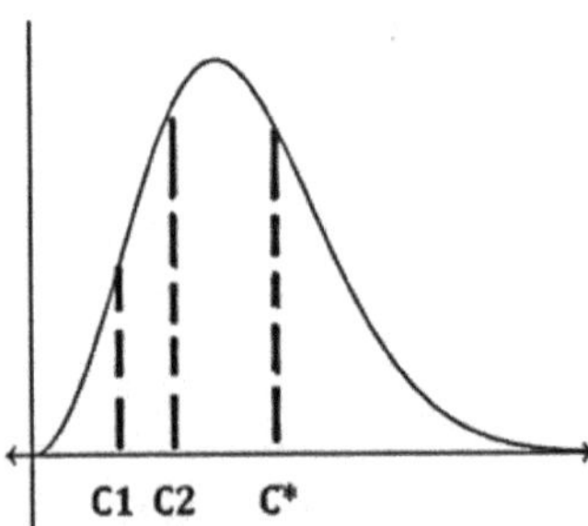

Figure 102: Conciousness

"Once again, each part of the distribution is dependent on all the other parts. This is because, at the physical level, we sense only by contrast, and if entities are removed from the segment of states between C_1 *and* C_2 in Figure 102 [left], other entities would soon develop those levels of consciousness because there is no awareness, or sensing ability, to keep them from occupying those states. Thus, it is with a broad spectrum of the distribution that an entity needs to meaningfully interact in order to enhance his/her consciousness. Again, one part of the system is dependent upon the others and collectively we form a species. The important thing to remember is that all members of the ensemble are part of the one – the whole system is ourselves. In part, I am you and you are me. That which I do reflects your growth and that what you do reflects mine. When we see something in our individual entity that we do not esteem, we cannot disown it and must find a suitable way to work on it until it is in harmony – for we are perfecting ourselves. And this we do by increasing the average consciousness C^* of our ensemble.

Both William Tiller and Michael Talbot, author of <u>The Holographic Universe</u>, published in 1991, write about paranormal activity.

"... Tiller thinks reality is similar to the "holodeck" on the television show Star Trek: The Next Generation. In the series, the holodeck is an environment in which occupants can call up a holographic simulation of literally any reality they desire, a lush forest, a bustling city. They can also change each simulation in any way they want, such as cause a lamp to materialize or make an unwanted table disappear. Tiller thinks the universe is also a kind of holodeck. Created by the "integration" of all living things. "we've created it as a vehicle of experience, and we've created the laws that govern it," he asserts. "And when we get to the frontiers of our understanding, we can in fact shift the laws so that we're also creating the physics as we go along." "

In physics, an electron volt (eV) is the amount of kinetic energy gained by a single electron accelerating from rest through an electric potential difference of one volt in vacuum. When used as a unit of energy, the numerical value of 1 eV in joules is equivalent to the numerical value of the charge of an electron in coulombs. Under the 2019 redefinition of the SI base units, this sets 1 eV equal to the exact value 10^{19} $1.602176634\times$ joules.

For close to 60 years, underground particle accelerators, some of them 500 miles long, have split the atom to see what was inside. The result has been a fairly large set of particles and sub particles.

Elementary particles are particles with no measurable internal structure; that is, it is unknown whether they are composed of other particles. They are the fundamental objects of quantum field theory. Many families and sub-families of elementary particles exist. Elementary particles are classified according to their spin. Fermions have half-integer spin while bosons have integer spin. All the particles of the Standard Model have been experimentally observed, recently including the Higgs boson in 2012. Many other hypothetical elementary particles, such as the graviton, have been proposed, but not observed experimentally. Other elementary categories of particles include leptons and quarks. Leptons consist of charged particles and neutral neutrinos.

Mathematical theories have predicted other particles, none of which have been observed experimentally. Next is a partial list.

Superpartner	Spin	superpartner of:	Name	Spin
chargino	$\frac{1}{2}$	charged bosons	axion	0
gluino	$\frac{1}{2}$	gluon	axino	$\frac{1}{2}$
gravitino	$\frac{1}{2}$	graviton	branon	?
Higgsino	$\frac{1}{2}$	Higgs boson	chameleon	0
neutralino	$\frac{1}{2}$	neutral bosons	dilaton	0
photino	$\frac{1}{2}$	photon	dilatino	$\frac{1}{2}$
sleptons	0	leptons	dual graviton	2
sneutrino	0	neutrino	graviphoton	1
squarks	0	quarks	graviscalar	0
wino, zino	$\frac{1}{2}$	W⁺ and Z⁰ bosons	inflaton	0
			magnetic photon	?
			majoron	0
			majorana fermion	$\frac{1}{2}$; $\frac{1}{2}$?...
			saxion	0
			X17 particle	?
			X and Y bosons	1
			W' and Z' bosons	1

The question becomes: Does Professor Tiller have a point? Are at least some of these particles wishful thinking? Many of these particles have a mass in the order of $10^{-36}\ kg$. A cyclotron or particle accelerator at best can measure to units of 10^{-23} and then only as a probability distribution.

Again, here are the decision making proportions.

$$(141) \qquad Formulative = \frac{a}{b} = \frac{c}{d}$$

$$(142) \qquad Relative = \frac{a}{b} = \frac{b}{c}$$

$$(143) \qquad Dimensional = \frac{a}{b} = \frac{b}{a+b}$$

The formulative proportion has 4 values that relate to each other relatively. The relationship of disproportionate facts is not science. The relative 3 term proportion has an observer that can marry two facts in a scientific way. The problem for Professor Tiller is that a passionate observer may see things in a wishful way to support their research. The dimensional proportion relates two facts in an intrinsic way by the $\sqrt{5}$ or possibly the Golden Ratio. The proportion is the only guaranteed truth.

Everyone has a DNA code that makes you human. The DNA tells the 20 amino acids what to do. It provides many proportions in your body that are dimensional; that is, they conform to a two term ratio. Many of your organs have a fractal structure.

While the figure on your right might give you vertigo, symbolically it represents a matrix of values whose elements grow with recursive iterations. In this example, a recursive matrix B representing a function with discrete values will be multiplied by a coefficient matrix A that represents a process acting on the function. The process matrix never changes.

Figure 103:
Expanding Squares

$$(144)\quad \textit{Kronecker product} = A \otimes B \qquad \textit{where}$$

- $A = process\ matrix$
- $B = function\ matrix$
- $A \otimes B =$
  $a\ tensor\ multiplication\ with\ special\ matrix\ properties$

Leopold Kronecker (1823-1891) was a German mathematician whose main contributions centered around the theory of equations. The matrix multiplication named after him results in a square matrix with special properties. In a way, it is a shorthand for regular matrix multiplications, has an easy way to find eigenvalues, and has a more complex commutative property. Our main concern is how the Kronecker product is used. To show a brief example, understand that each coefficient in A multiplies the entire matrix B.

$$(144) \qquad B_2 = \begin{bmatrix} a_{11}B & a_{12}B \\ a_{21}B & a_{22}B \end{bmatrix}$$

For example, given

$$(145) \qquad A = \begin{bmatrix} 1 & 0 \\ 1 & 1 \end{bmatrix} \text{ and } B = A,$$

$$(146) \qquad A \otimes B = \begin{bmatrix} 1 & 0 \\ 1 & 1 \end{bmatrix} \otimes \begin{bmatrix} 1 & 0 \\ 1 & 1 \end{bmatrix} = \begin{bmatrix} 1 & 0 & 0 & 0 \\ 1 & 1 & 0 & 0 \\ 1 & 0 & 1 & 0 \\ 1 & 1 & 1 & 1 \end{bmatrix} = B_1$$

If the zeros represent blank spaces, are you beginning to see a pattern? Notice that the 2x2 matrix has become a 4x4 matrix. If we iterate again, the result will be an 8x8 matrix. For each iteration, the matrix grows by a 2 to 1 ratio, the ratio of the musical octave. After 10 iterations, the result is shown below.

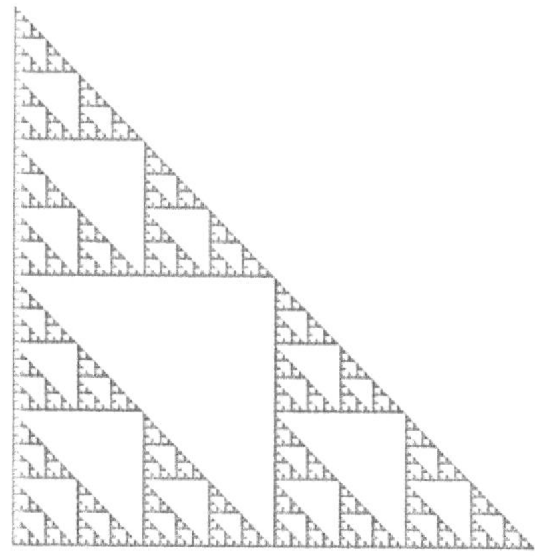

Figure 104: Kronecker Product Sierpinski Triangle

This almost magical result is a right triangle version of a Sierpinski Gasket, a basic fractal example. The software code that iterated the Kronecker product came from a 2017 internet article entitled *Generating Kronecker Product Based Fractals*, Dr. A. E. Voevudko, architect. There are many ways to generate fractals besides the Kronecker product, but this method demonstrates how a repeated process yields results. Also, the ratio increase in the number of elements used to create this self-similar result is made up of whole numbers. Do these results represent reality? As a student example -- no. As a demonstration of a process of Nature, the answer is definitely yes. What follows are more examples of Kronecker products.

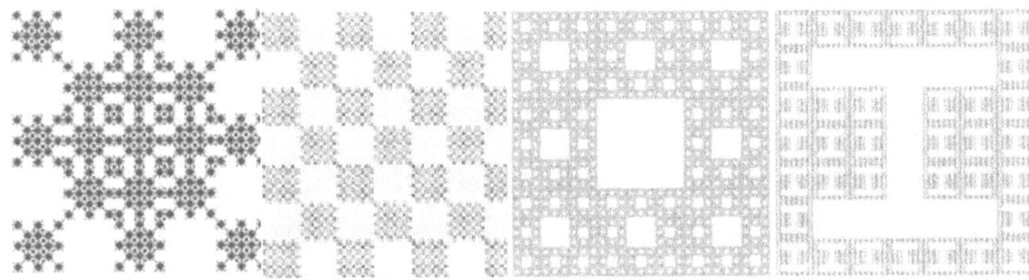

Figure 105: Kronecker Product Fractal Examples

Chapter 17: The New World

Whether or not you believe that a better understanding of science leads to a new world depends on your perspective. If you know the laws of nature, then do these laws translate to human behavior? If science is regulated by random chance, if particles move without memory, if electrons stay or leave their orbits on the flip of a coin, then what chance have you got as a human being? With no memory, if you step in a mud puddle while crossing the street, there is a finite probability that you will step in the puddle again. Heisenberg believes in the uncertainty of motion. Einstein believes that God does not roll the dice.

The English translation of Werner Heisenberg's book, <u>Physics and Beyond: encounters and conversations</u>, occurred in 1972. In January 1933, Heisenberg and a few colleagues spent a few days vacationing at a ski chalet in Bayrichzell, Gemany. With his magnificent memory, Heisenberg (WH) recalls a conversation with Neils Bohr (NB).

WH Isn't it odd that, throughout this discussion, no one should have mentioned quantum theory? We behave as if electrically charged particles were an object like an electrically charged water droplet, or like a pithball in an old electroscope. We quite unthinkingly use the concepts of classical physics, as if we had never heard of the limitations of these concepts and of these uncertainty relations. Isn't that bound to lead to errors?

NB No, certainly not After all, it is almost the essence of the experiments that the observations can be described with the concepts of classical physics. That is the whole paradox of quantum theory. On the one hand, we establish laws that differ from those of classical physics;

on the other, we apply the concepts of classical physics quite unreservedly whenever we make observations, or take measurements or photographs. And we have to do just that because, when all is said and done, we are forced to use language, if we are to communicate our results to other people. A measuring instrument is a measuring instrument only when the observations it yields enable us to arrive at unequivocal conclusions about the phenomenon under observation, only when a strict casual connection can be presumed to exist. Yet when it comes to the theoretical description of an atomic phenomenon, we must make a distinction between the phenomenon or the observer and his apparatus. The demarcation line may be subject to choice, but on the observer's side of the split we are forced to use the language of classical physics, simply because we have no other language in which to express the results. We know that the concepts of this language are imprecise, that they have a limited area of application, but we have no other language, and, after all, it does help us to grasp the phenomenon at least indirectly.

WH　Isn't it possible that once we have understood quantum theory even better, we might be able to dispense with the classical concepts and use a new language to speak far more accurately about the atomic phenomena than we can today?

NB　You are misunderstanding the problem. Science is the observation of phenomena and the communication of results to others, who must check them. Only when we have agreed about what is happening objectively, or on what happens regularly, do we have a basis for understanding. And this whole process of communication and observation proceeds by means of the concepts of classical physics. ... It is one of the presuppositions of science that we speak of measurements in a language that has basically the same structure as the one in which we speak of everyday experience. We have learned that this language is an inadequate means of communication and orientation, but nevertheless the presupposition of all science.

In today's world, virtually every science as well as every occupation has a language that is hardly decipherable by the "language of everyday experience". In electrical engineering, the units of frequency were changed from cycles per second to "hertz" in 1963. Heinrich Hertz (1857 – 1894), German physicist proved the existence of electromagnetic waves predicted by James Maxwell. Certainly, the honor of the frequency unit may have been well deserved but it does not help communication with the public.

From Chapter 3 and after eight years of checking, the Cern scientists released their findings that a particle acceleration of neutrino beams exceeded the speed of light by 60 nanoseconds. The next year they retracted their findings because they found an unconnected wire. With 500 miles of cyclotron magnets and wires, the complexity even for the scientists was not realizable. Popular science books like this will be harder to write because of the snowflakes of technology.

Toward the end of his life, Albert Einstein shared his retrospective views about his experiences. In 1950, Einstein wrote a letter to M. Berkowitz. The letter was quoted in a book by William Hermanns entitled <u>Einstein and the Poet, In Search of the Cosmic Man</u> published in 1983.

- I believe that the most important question facing humanity is, 'Is the universe a friendly place?' This is the first and most basic question people must ask themselves.

- For if we decide that the universe is an unfriendly place, then we will use our technology, our scientific discoveries and our natural resources to achieve safety and power by creating bigger walls to keep out the unfriendliness and bigger weapons to destroy all that which is unfriendly and I believe that we are getting to a place where technology is powerful enough that we may either completely isolate ourselves or destroy ourselves in this process.

- If we decide that the universe is neither friendly or unfriendly and that God essentially 'is playing dice with the universe,' then we are simply victims of the random toss and our lives have no real purpose or meaning.

- God does not play dice with the Universe.

After over 300 technical papers and a dozen books, Einstein's ultimate goal was to describe a universal equation. He was not introduced to chaos theory that describes how living things grow. When we have elections, the populous bifurcates and decides which direction to head. Our world is fractal. The Universe is fractal. It makes decisions based on whole number ratios. When there are rapid gains in the stock market, usually there is a 50 percent retreat so that the logic of this activity can be reexamined. The planets orbit around the sun in whole number ratios with each other.

The ancients solved non-linear problems without the aid of calculus. They observed the calendar, created the zodiac, built the pyramids, discovered time, understood the cycles of the moon. Their mathematics included the arithmetic mean, harmonic mean, and geometric mean. They observed balance in motion.

The Constitution of the United States was founded on the principles of balance in motion. The three branches of government – legislative, executive, and judicial – act as checks and balances against moving too fast or too slow. In theory, their decisions must be in harmony with the will of the people. The legislative branch represents the arithmetic mean, the executive branch represents the geometric mean, and the judicial branch represents the harmonic mean.

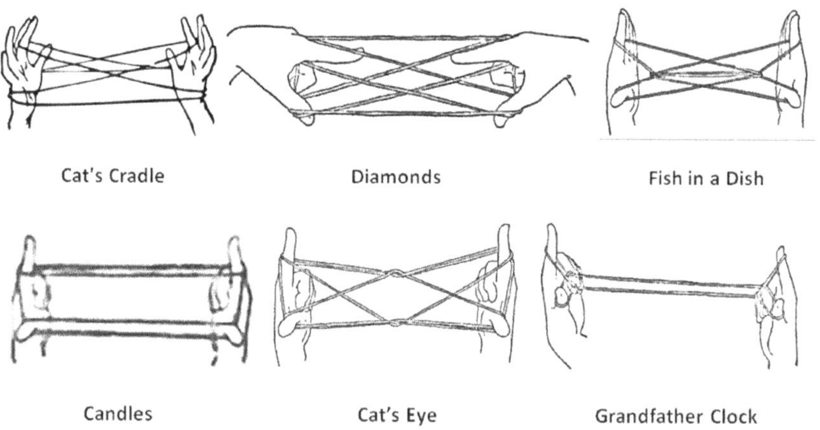

Figure 106: Cat's Cradle String Game

The ancient children's game of "Cat's Cradle" requires a length of string and is played all over the world. It can be played singularly, with two players, and with a group of players. The string is looped around your fingers and stretched to form a pattern. If two players are competitive, then a player will tangle its fingers inside the opposing player's pattern, stretch its fingers, and hope to create a new pattern. If the player is unsuccessful or if the new pattern is not pleasing to the players' sight, then the player loses. Otherwise the game continues.

Without trying, the child is experiencing order, pattern, symmetry, proportion, harmony; all synonyms for life. This is what the Universe is trying to tell you. If a person is in harmony with his/her self, with their neighbors and friends, with their environment, and with the world, then even if you are not successful by some internal measure, you will create success in others. These pathways are the roadways to the creation of the Universe.

Perhaps the most famous author of scientific knowledge was the astronomer Carl Sagen. He wrote over 20 popular science books from 1960 to 1990.

The scientific way of thinking is at once imaginative and disciplined. This is central to its success. Science invites us to let the facts in, even when they don't conform to our preconceptions. It counsels us to carry alternative hypotheses in our heads and see which best fit the facts. It urges on us a delicate balance between no-holds-barred openness to new ideas, however heretical, and the most rigorous skeptical scrutiny of everything—new ideas and established wisdom.

We all gain knowledge by challenging accepted reasoning and providing new ideas. You are in charge of your personal destiny. May you continue your curious journey to expand our friendly Universe.

Appendix A: Conversation between Albert Einstein and Werner Heisenberg

In the spring of 1926 just outside the University of Berlin, two of the most famous scientists of the 20th Century went for a walk. The So1vay Congress was in session to discuss the new theory of quantum mechanics. This was the first time Werner Heisenberg had met Albert Einstein to have any serious discussion. In his 1932 book, <u>Physics and Beyond: Encounters and Conversations</u>, Werner Heisenberg recalls their conversation.

Among the translations from German, Harper & Sons published their hardcover book with the same title in 1961 and their softcover 1972. The excerpts from these translations are quotations only. As such, the left margin uses the abbreviation "AE" for Albert Einstein and "WH" for Werner Heisenberg.

AE What you have told us sounds extremely strange. You assume the existence of electrons inside the atom, and you are probably quite right to do so. But you refuse to consider their orbits, even though we can observe electron tracks in a cloud chamber. I should very much like to hear more about your reasons for making such strange assumptions.

WH We cannot observe electron orbits inside the atom, but the radiation which an atom emits during discharges enables us to deduce the frequencies and corresponding amplitudes of its electrons. After all even in the older physics wave numbers and amplitudes could be considered substitutes for electron orbits. Now, since a good theory must be based on directly observable magnitudes, I thought it more fitting to restrict myself to these, treating them, as it were, as representatives of the electron orbits.

AE   But you don't seriously believe that none but observable magnitudes must go into a physical theory?

WH   Isn't that precisely what you have done with relativity? After all, you did stress the fact that it is impermissible to speak of absolute time, simply because absolute time cannot be observed; that only clock readings, be it in the moving reference system or the system at rest, are relevant to the determination of time.

AE   Possibly I did use this kind of reasoning, but it is nonsense all the same. Perhaps I could put it more diplomatically by saying that it could be more heuristically useful to keep in mind what one has actually observed. But on principle, it is quite wrong to try founding a theory on observable magnitudes alone. In reality the very opposite happens. It is the theory that decides what we can observe. You must appreciate that observation is a very complicated process. The phenomenon under observation produces certain events in our measuring apparatus. As a result, further processes take place in the apparatus, which eventually and by complicated paths produce sense impressions and help us to fix the events in our consciousness. Along this whole path – from the phenomenon to the fixation in our consciousness – we must be able to tell how nature functions, must know the natural laws at least in practical terms, before we can claim to observe anything at all. Only theory, that is, knowledge of natural laws, enables us to deduce the underlying phenomena from our sense impressions. When we can claim that we can observe something new, we ought really to be saying that, although we are about to formulate new natural laws that do not agree with the old ones, we nevertheless assume that the existing laws – covering the whole path from the phenomenon to our consciousness – function in such a way that we can rely on them and speak of 'observations.'

AE   In the theory of relativity, for instance, we presuppose that, even in the moving reference system, the light system travelling from the clock to the observer's eye behave more or less as we have always expected them to behave. And in your theory, you quite obviously assume that the whole mechanism of light transmission from the vibrating atom to the spectroscope or to the eye works just as one always supposed it does, that is, essentially according to Maxwell's laws. If that were no longer the case, you could not possibly observe any of the magnitudes you call observable. Your claim that you are observing none but observable magnitudes is therefore an assumption about a property of the theory that you are trying to formulate. You are, in fact, assuming that your theory does not clash with the old description of radiation phenomena in the essential points. You may well be right, of course, but you cannot be certain.

WH   The idea that a theory is no more than a condensation of observations in accordance with the principle of thought economy surely goes back to Mach, and it has, in fact, been said that your relativity theory makes decisive use of Machian concepts. But what you have just told me seems to indicate the very opposite. What am I to make of all this, or rather what do you yourself think about it?

AE   It's a very long story, but we can go into it, if you like. Mach's concept of thought economy probably contains part of the truth, but strikes me as being just a bit too trivial. Let me first of all produce a few arguments in its favor. We obviously grasp the world by way of our senses. Even when small children learn to speak and to think, they do so by recognizing the possibility of describing highly complicated but somehow related sense impressions with a single word, for instance, the word 'ball.' They learn it from adults and get the satisfaction that they can make themselves understood. In other words, we may argue that the formation of the word, and hense of the concept, 'ball' is kind of a thought economy enabling the child to combine very complicated sense impressions in a simple way. Much does not even enter into the question which mental or physical predispositions must be satisfied in man – or the small child – before the process of communication can be initiated. With animals, the process works considerably less effectively, as everyone knows, but we shan't talk about that now. Now Mach also thinks that the formation of scientific theories, however complex, takes place in a similar way. We try to order the phenomena, to reduce them to a simple form, until we can describe what may be a large number of them with the aid of a few simple concepts.

WH   All this sounds very reasonable, but we must nevertheless ask ourselves in what sense the principle of mental economy is being applied here. Are we thinking of psychological or logical economy, or, again, are we dealing with subjective or the objective side of the phenomena? When the child forms the concept 'ball', does he introduce a purely psychological simplification in concept, or does this ball really exist? Mach would probably answer that the two statements express one and the same fact. But he would be quite wrong to do so. To begin with, the assertion 'The ball really exists' also contains a number of statements about possible sense impressions that may exist in the future. New future possibilities and expectations make up a very important part of our reality, and must not be simply forgotten. Moreover, we ought to remember the inferring concept and things from sense impressions is one of the basic presuppositions of all our thought. Hence, if we wanted to speak of nothing but sense impression, we should have to rid ourselves of our language and thought. In other words, Mach rather neglects the fact that the world realty exists, that our sense impressions are based on something objective.

AE   I have no wish to appear as an advocate of a naive form of realism: I know that these are very difficult questions, but then I consider Mach's concept of observation also much too naïve. He pretends that we know perfectly well what the word 'observe' means, and thinks that this exempts him from having to discriminate between 'objective' and 'subjective' phenomena. No wonder the principle has an suspiciously commercial a name: 'thought economy.' His idea of simplicity is much too subjective for me. In reality, the simplicity of the natural laws is an objective fact as well, and the correct conceptual scheme must balance the subjective side of this simplicity with the objective. But this is a very difficult task. Let us rather return to your lecture.

AE   I have a strong suspicion that, precisely because of the problems that we have just been discussing, your theory will one day get you into hot water. I should like to explain this in greater detail. When it comes to observation, you behave as if everything can be left as it was, as if you could use the old descriptive language. In that case, however, you will also have to say: in a cloud chamber we can observe the path of the electrons. At the same time, you claim that there are no electron paths inside the atom. This is obvious nonsense, for you cannot possibly get rid of the path simply by restricting the space in which the electron moves.

WH   For the time being, we have no idea in what language we must speak about processes inside the atom. True, we have a mathematical language, that is, a mathematical scheme for determining the stationary states of the atom or the transition probabilities from one state to another, but we do not know – at least not in general – how the language is related to that of classical physics. And, of course, we need this connection if we are to apply this theory to experiments in the first place. For when it comes to experiments, we inevitably speak in the traditional language. Hence I cannot really claim that we have 'understood' quantum mechanics. I assume that the mathematical scheme works, but no link with the traditional language has been established so far. And until that has been done, we cannot hope to speak of the path of the electron in the cloud chamber without inner contradictions. Hence it is probably much too early to solve the difficulties you have mentioned.

AE   Very well, I will accept that. We shall talk about it again in a few years' time. But perhaps I may put another question to you. Quantum theory as you have expounded it in your lecture has two distinct faces. On the one hand, as Bohr himself has rightly stressed, it explains the stability of the atom; it causes the same forms to appear time and again. On the other hand, it explains that strange discontinuity or inconsistency of nature which we observe quite clearly when we watch flashes of light on a scintillation screen. These two aspects are obviously connected. In your quantum mechanics you will have to take both into account, for instance when you speak of the emission of light by atoms. You can calculate the discrete energy values of the stationary states. Your theory can then account for the stability of certain forms that cannot merge continuously into one another, but must differ by finite amounts and seem capable of permanent reformation. But what happens during the emission of light? As you know, I suggested that, when an atom drops suddenly from one stationary energy value to the next, it emits the energy difference as an energy packet, a so called light quantum. In that case, we have a particularly clear example of discontinuity. But you think that my conception is correct? Or can you describe the transmission from one stationary state to another in a more precise way?

WH   Bohr has taught me that one cannot describe the

process by means of the traditional concepts by, i.e., as a process in time and space. With that, of course, we have said very little, no more, in fact, than that we do not know. Whether or not I should believe in light quanta, I cannot say at this stage. Radiation quite obviously involves discontinuous elements to which you refer as light quanta. On the other hand, there is a continuous element, which appears, for instance, in interference phenomena, and which is much more simply described by the wave theory of light. But you are of course quite right to ask whether quantum mechanics has anything new to say on these terribly difficult problems. I believe that we may at least hope that we will one day.

WH   I could, for instance, imagine that we should obtain an interesting answer if we considered the energy fluctuations of an atom during reactions with other atoms or with the radiation field. If the energy should change discontinuously, as we express from your theory of light quanta, then the fluctuation, or, in more precise mathematical terms, the mean square fluctuation, would be greater than if the energy changed continuously. I am inclined to believe that quantum mechanics would lead to the greater value; and so establish the discontinuity. On the other hand, the continuous element, which appears in interference experiments, must also be taken into account. Perhaps one must imagine the transitions from one stationary state to the next as so many fade outs in a film. The change is not sudden – one picture gradually fades while the next comes into focus so that, for a time, both pictures become confused and one does not know which is which. Similarly, there may well be an intermediate state in which we cannot tell whether an atom is in the upper or lower state.

AE   You are moving on very thin ice, for you are suddenly speaking of what we know about nature and no longer about what nature really does. In science we might to be concerned solely with what nature does. It might very well be that you and I know quite different things about nature. But who would be interested in that? Perhaps you and I alone. To everyone else it is a matter of complete indifference. In other words, if your theory is right, you will have to tell me sooner or later what the atom does when it passes from one stationary state to the next.

WH Perhaps. But it seems to me that you are using a language a little to strictly. Still, I do admit that everything I might now say may sound like a cheap exercise. So let's wait and see how atomic theory develops.

AE How can you really have so much faith in your theory when so many crucial problems remain completely unsolved?

WH I believe, just like you, that the simplicity of natural laws has an objective character, that it is not just the result of thought economy. If nature leads us to mathematical forms of great simplicity and beauty – by forms I am referring to coherent systems of hypotheses, axioms, etc. – to forms that no one has previously encountered, we cannot help thinking that they are 'true,' that they reveal a genuine feature of nature. It may be that these forms also cover our subjective relationship to nature, that they reflect elements of our own thought economy. But the mere fact that we could never have arrived at these forms by ourselves, that they were revealed to us by nature, suggests strongly that they must be part of reality itself, not just of our thoughts about reality.

WH You may object that by speaking of simplicity and beauty I am introducing aesthetic criteria of truth, and I frankly admit that I am strongly attracted by the simplicity and beauty of the mathematical schemes with which nature presents us. You must have felt this, too; the almost frightening simplicity and wholeness of the relationships which nature suddenly spreads out before us and for which none of us was in the least prepared. And this feeling is something completely different from the joy we feel when we have done a set task particularly well. That is one reason why I hope that the problems we have been discussing will be solved in one way or another. In the present case, the simplicity of the mathematical scheme has the further consequence that it ought to be possible to think up many experiments whose results can be predicted from the theory. And if the actual experiments should bear out the predictions, there is little doubt but that the theory reflects nature accurately in this particular realm.

AE   Control by experiment, is, of course an essential prerequisite of the validity of any theory. But one can't possibly test everything. That is why I am so interested in your remarks about simplicity. Still, I should never claim that I really understood what is meant by the simplicity of the natural laws.

Appendix B: List of Illustration Attributes

1. *Einstein Photon Box.* Proposed Experiment at the Solvay Congress in 1930. Retrieved from https://commons. wikimedia.org/wiki/File:Einstein%27s_light_box.svg, attrib-uted to Wikipedia commons and Prokaryotic Caspase Homolog, accessed 11/12/20.

2. *Schrodinger's Cat.* Famous thought analogy by Erwin Schrodinger. Schrodinger's cat in a closed, booby-trapped box with a random quantum trigger could be simultaneously dead and alive. From here, the only way to know whether the cat is dead or alive is to open the box and look. By opening the box, you risk killing the cat. Retrieved from https: //thumbs. dreamstime.com/z/schr%C3%B6dingers-cat-schrodingers-closed-booby-trapped-box-random-quantum-trigger-could-be-simultaneously-dead-alive-40711742.jpg, attributed to and purchased from dreamstime.com, accessed 11/16/20.

3. *Plato Geometry.* Explanation of figures from Timeas. Retrieved from href='https://www.freepik.com/vectors/background'> Background vector created by vectorpouch, attributed to www.freepik.com, accessed 11/12/20.

4. *Plato Proportions.* Geometry proportions illustrating Plato's description from Timeas. Original illustration.

5. *Third Order Polynomial.* Illustration of the double bending curve crossing the X-axis. Original drawing.

6. *Paper Sizes.* Standard paper sizes traced back to ancient geometry. Retrieved from https://thumbs.dreamstime.com/z/series-paper-sizes-labels-simple-flat-vector-illustration-171788333.jpg, attributed to and purchased from dreamstime.com, accessed 11/15/20.

7. *Apollo and the Sun.* An 18th Century fresco located in the Gottweig Abbey in Krems, Austria. The painting decorates the imperial staircase and is considered a masterpiece of Baroque architecture. It was completed by Paul Troger and depicts the Holy Roman Emperor Charles VI in the form of the Greek God Apollo, riding bare-chested in his sun chariot wearing his powdered wig. Retrieved from https://thumbs.dreamstime.com/z/g%C3% B6ttweig-abbey-fresco-th-century-located-krems-austria-painting-dec-orates-imperial-staircase-considered-129873664.jpg, attributed to and purchased from dreamstime.com, 11/16/20.

8. *Michelson-Morley Experiment.* Legendary mirror experiment to measure the speed of light as the Earth rotates. Retrieved by https://www.writework.com/essay/michelson-and-morey-experiment, attributed to WriteWork contributors The Michelson and Morey Experiment, accessed 11/ 17/20.

9. *Marett Reconstruction of Silvertooth Experiment.* Photograph of first lasar-mirror table replicating the Silvertooth experiment. Retrieved from Doug Marett A Replication of the Silvertooth Experiment, attributed to Doug Marett Sky Hunt Scientific, accessed 11/27/20.

10. *Schematic of Marett's Experimental Table.* A schematic showing the mirrors, components, and light paths of the experiment. Retrieved from Doug Marett A Replication of the Silvertooth Experiment, attributed to Doug Marett Sky Hunt Scientific, accessed 11/27/20.

11. *Oscilloscope Results of Marett Experiment.* A window with two waveforms separated in time. Original illustration.

12. *Rosette Nebula.* This infrared image from NASA's Spitzer Space Telescope shows the Rosette Nebula, a pretty star-forming region more than 5,000 light-years away in the constellation Monoceros. In optical light, the nebula looks like a rosebud, or the "rosette" adornments that date back to antiquity. Retrieved from https://imagecache.jpl.nasa.gov/images/ /pia13126-640-640x 350.jpg, attributed to NASA/JPL-Caltech/Univ. of Ariz., accessed 11/27/20.

13. *Da Vinci's Vitruvian Man.* Leonardo Da Vinci from 1492. Retrieved from https://thumbs.dreamstime.com/z/da-vinci-s-vit-ruvian-man-6211519.jpg, attributed to and purchased from dreamstime.com, accessed 11/16/20.

14. *Pythagorean Universal Constants.* A triangle showing the Pythagorean relationship between e, $\varnothing$, and π. Original illustration.

15. *Experimental Results.* The graphs of results from compounding e, $\varnothing$, and π 360 times. Original illustrations.

16. *Musical Overtone Series.* The notes generated as part of the overtone series of a fundamental tone. Original illustration.

17. *Escher's Ascending and Descending.* Escher graphic of people walking either up the stairs or down. https://upload.wikimedia.org/wikipedia/en/6/66/Ascending_and_Descending.jpg, attributed to Wikipedia Ascending and Descending, accessed 11/17/20.

18. *NASA First Black Hole Image (2019).* Visual proof that black holes exist. Retrieved by https://www.nasa.gov/sites/default/files/styles/full_width/public/thumbnails/image/blackhole.png?itok=THJrwcHP, attributed to NASA, accessed 11/17/20.

19. *Space-Time Spherical Coordinates.* A sphere illustrating that space-time straight lines are curved in space because of gravity. Retrieved from https://upload.wikimedia.org/wikipedia/commons/thumb/a/a2/Kugelkoord-lokale-Basis-s.svg/800px-Kugelkoord-lokale-Basis-s.svg.png, attributed to Wikimedia commons, accessed 11/27/20.

20. *Cassiopeia – A Supernova Remnant.* Retrieved by https://solarsystem.nasa.gov/system/resources/detail_files/822_casa_elements_detail.jpg, attributed to NASA, retrieved 11/17/20.

21. *Rocket Launch.* Space rocket takeoff at the Cape. Retrieved from https://images.unsplash.com/photo-1517976487492-5750f3195933?ixlib=rb-1.2.1&ixid=eyJhcHBfaWQiOjEyMDd9&auto=format&fit=crop&w=1500&q=80, attributed to NASA, 11/14/20.

22. *Singularities in Space-Time.* Roger Penrose's Black Hole singularity derived from general relativity. Retrieved from https:// en.wikipedia.org/wiki/Penrose_diagram, attributed to Wikipedia, accessed 11/13/20.

23. *Gravity Slingshot Effect.* Drawing showing acceleration around a planet due to its gravity. Received from http://sciphile.org/lessons/stacked-ball-drop-lessons-conservation-energy-and-momentum, attributed to creative commons, accessed 11/14/20.

24. *Godel Dust Cones.* Kurt Godel's theory of black hole dust cones derived from general relativity. Retrieved from https:// en.wiki-pedia.org/wiki/G%C3%B6del_metric, attributed to Wikipedia Godel Metric, accessed 11/13/20.

25. *Harold Hurst Treading a Nile River Outlet.* Received from article Harold Edwin Hurst: the Nile and Egypt, past and future, https://doi.org/10.1080/02626667.2015.1019508, attributed to John Sutcliffe,Stephen Hurst,Ayman G. Awadallah,Emma Brown and Khaled Hamed, pages 1557-1570, accessed 11/17/20.

26. *Sierpinski Triangle Panel.* Fractal geometry exercise using triangles that progressively divides into smaller triangles. Original illustration.

27. *Electron Etching of Sierpinski Triangle.* Electron etching of Sierpinski triangle bounded on a copper surface. Retrieved from https://doi.org/10.1038/s41567-018-0328-0, attributed to Kempkes, S.N., Slot, M.R., Freeney, S.E. et al. Design and characterization of electrons in a fractal geometry, Nature Phys 15, 127–131 (2019), accessed 11/27/19.

28. *Koch Snowflake Panel.* Fractal geometry exercise using triangles that progressively divide into smaller triangles while adding area to the side of a square. Retrieved from https://thumbs. dreamstime.com/z/koch-snowflake-construction-fractal-geo-metry-exercise-using-triangles-progressively-divides-smaller-white-black-160870568.jpg, attributed to and purchased from dreamstime.com.

29. *Nile River Water Flow Chart.* Over 137 years, the cubic kilo-meter water flow per year. Original design.

30. *Changing Hurst Exponents for a Stock Chart.* Comparing the rise and fall of stock price with its measured Hurst Exponent. Original simulation.

31. *The Planets.* Planets of solar system orbiting around sun on cosmic background with meteorites and asteroids, infographic illustration for school education or space exploration. Retrieved from https://thumbs.dreamstime.com/z/planets-solar-system-orbit-ing-around-sun-cartoon-vector-cosmic-back-ground-meteorites-aster-oids-infographic-153811407.jpg, attri-buted to and purchased from dreamstime.com, accessed 11/16/20.

32. *The Overtone Series Revisited.* Musical whole notes showing the overtone pitches from the fundamental tone. An original illustration.

33. *Pythagoras Music of the Spheres.* Pythagoras conceived the uni-verse to be an immense monochord, with its single string connected at its upper end to absolute spirit and at its lower end to absolute matter — in other words, a cord stretched between heaven and earth. Counting inward from the circumference of the heavens, Pythagoras, according to some authorities, divided the universe into nine parts; according to others, into twelve parts. The twelvefold system was as follows: The first division was called the empyrean, or the sphere of the fixed stars, and was the dwelling place of the immortals. The second to twelfth divisions were (in order) the spheres of Saturn, Jupiter, Mars, the sun, Venus, Mercury, and the moon, and fire, air, water, and earth. This arrangement of the seven planets (the sun and moon being regarded as planets in the old astronomy) is identical with the candlestick symbolism of Jewish philosophy — the sun in the center as the main stem with three planets on either side of it. Retrieved from https://against-the-day.pynchon-wiki.com/wiki/images/thumb/4/47/Harmonies-of-the-spheres.jpg/225px-Harmonies-of-the-spheres.jpg, attributed to The Secret Teach-ings of All Ages by Manly P. Hall (1928), accessed 11/27/20.

34. *Kepler Harmonicis Mundi Music (1619).* Retrieved from https:// upload.wikimedia.org/wikipedia/commons/thumb/d/db/Pla netary_Musical_Scales_from_Harmony_of_the_Worlds.jpg/80 0px-Planetary_Musical_Scales_from_Harmony_of_the_ Worlds.jpg, attributed to Wikipedia Harmonicis Mundi, accessed 11/16/20.

35. *Milky Way Galaxy.* The Milky Way spiraling like a disk with curved tentacles. Retrieved from https://thumbs.dreamstime. com/z/spiral-galaxy-milky-way-d-image-35274059.jpg, attri- buted to and purchased from dreamstime, accessed 11/18/20.

36. *Solar Wind.* Ionized particles from the sun that create a mag- netic field. Retrieved from https://thumbs.dreamstime. com/ z/earths-magnetic-field-earth-solar-wind-flow-particles 884846 89.jpg, attributed to and purchased from dreams-time.com, accessed 11/16/20.

37. *Northern Lights.* Northern lights take place on Earth as well as other planets. Jupiter lights retrieved from https://apod. nasa.gov/apod/image/1009/AuroraPrelude_takasaka.jpghttps: //www.jpl.nasa.gov/spaceimages/images/largesize/PIA21938_ hires.jpg, attributed to NASA (9/6/17), accessed 11/14/20.

38. *Curl Right Hand Rule.* Curl right hand rule detecting direction of the induced current flow by direction of magnetic field. Recieved from https://thumbs.dreamstime.com/z/curl-right- hand-rule-vec-tor-illustration-example-diagram-detecting-dir- ection-induced-cur-rent-flow-magnetic-field-physics-173736679. jpg, attributed to dreams time.com, accessed 11/16/20.

39. *Resonance.* The gaps in the Asteroid Belt are due to the gravity resonances of Jupiter. Retrieved from https://upload.wiki- media.org/wikipedia/commons/thumb/0/07/Resonance.PNG /440px-Resonance.PNG, attributed to Wikipedia Reson-ance, accessed 11/28/20.

40. *Moon Orbits around Jupiter.* The orbits of Jupiter's three closest and most famous moons have geometric ratios that allow the moons to align every 8 years. Retrieved from https://upload.wiki-media.org/wikipedia/commons/e/e5/Galilean_moon_Laplace_resonance_animation_2.gif, attributed to Wikipedia commons, 11/27, 2020.

41. *Densities of Asteroid Belt.* Kirkwood Gaps show the effect of Jupiter's gravity on the Asteroid Belt. Retrieved from https://upload.wikimedia.org/wikipedia/commons/thumb/d/d3/Kirkwood_Gaps.svg/800px-Kirkwood_Gaps.svg.png, attributed to Wikipedia Asteroid belt , accessed 11/14/20.

42. *Magellan Galaxies.* ESO's Very Large Telescope has captured a detailed view of a star forming region in the Large Magellanic Cloud — one of the Milky Way's satellite galaxies. This sharp image reveals two glowing clouds of gas. NGC 2014 (right) is irregularly shaped and red and its neighbor, NGC 2020, is round and blue. These odd and very different forms were both sculpted by powerful stellar winds from extremely hot newborn stars that also radiate into the gas, causing it to glow brightly. Retrieved from https:// upload.wikimedia.org/wikipedia/commons/thumb/9/9c/Two_very_different_glow-ing_gas_clouds_in_the_Large_Magellanic_Cloud.jpg/800px-Two_very_different_glowing_gas_clouds_in_the_Large_ Mag-ellanic_Cloud.jpg, attributed to Wikipedia commons Two very different glowing gas clouds in_the Large Magellanic Cloud, accessed 11/28/20.

43. *Stars in the Night Sky.* Contrasting star clusters exploring the brightness of the night sky. Retrieved from https://apod.nasa.gov/apod/image/0708/M67_Greg_Noel800.jpg, and retrieved from https://www.nasa.gov/sites/default/files/ styles/ full_width_feature/public/thumbnails/image/potw 1643a.jpg, attributed to NASA, accessed 11/28/20.

44. *Big Bang through Time.* The expansion of the universe from the Big Bang to the present. Retrieved from https://thumbs. dreamstime.com/z/big-bang-expansion-universe-to-present-digital-illustration-63971494.jpg , attributed to and purchased from dreamstime.com, accessed 11/16/20.

45. *Wormhole.* Illustration of wormhole; curvature of space. Retrieve from https://thumbs.dreamstime.com/z/illustration -wormhole-d-model-grid-124485433.jpg, and retrieved from https://apod. nasa.gov/apod/image/0708/M67_GregNoel.jpg, attributed to and purchased from dreamstime.com and attributed to NASA, accessed 11/18/20.

46. *Quasar.* Quasar illustration. Retrieved from https:// www.nasa. gov/sites/default/files/thumbnails/image/stsci-h-2010a-d-1280 x720.png, attributed to NASA, accessed 11/18/20.

47. *Rabbits Verses Foxes.* Predator-Prey curve. Retrieved from https: //upload.wikimedia.org/wikipedia/commons/b/b6/Predator_ prey_curve.png, znd retrieved from https://thumbs. dreamstime.com/z/rabbit-fox-sly-red-scared-gray-85604101.jpg, attributed to Wikipedia Commons and attributed to and purchased from dreamstime.com, accessed 11/18/20.

48. *Brusselator Diagram.* The Brusselator is a theoretical model for a type of autocatalytic reaction possibly resulting in new particles and energetic life. Original illustration.

49. *Belousov-Zhabotinskii Chemical Reaction.* Generation of patterns in a liquid container. Retrieved from https://thumbs. dreams-time.com/z/belousov-zhabotinsky-reaction-chemical-reactions-abstract-representation-91457414.jpg, attributed to and purchased from dreamstime.com, accessed 11/16/20.

50. *Life Cycle of Stars.* Graphic history of stars. Received from https://imagine.gsfc.nasa.gov/Images/objects/stars_lifecycle_ full.jpg, attributed to NASA, accessed 11/18/20.

51. *Galaxy Count over 4 Billion Light-Years.* Chart showing statistically regenerated data from Paul LaViolette's 1995 book <u>Beyond the Big Bang</u>. Original diagram.

52. *Excel Calculation of the Fractal Density of Stars.* A snapshot of an Excel worksheet showing the work required to calculate a fractal dimension. Original chart.

53. *Platonic solids.* Nature's only Euclidian shapes for solids. Retrieved from https://thumbs.dreamstime.com/z/blue-platonic-solids-wireframe-models-173951510.jpg, attributed to and purchased from dreamstime.com, accessed 11/16/20.

54. *Fractal Nature of Lungs.* Lung diagram with fractal branches. Received from https://upload.wikimedia.org/wikipedia/ commons/thumb/a/a1/Lungs_diagram_detailed.svg/250px-Lungs _diagram_detailed.svg.png, attributed to Wikipedia Commons, accessed 11/18/20.

55. *Fractal Tree.* A simple exercise resulting in a symmetrical tree illustration. Original illustration.

56. *Fractal Dimension of British Coastline by Box Count.* Three drawings of British coastline with graphic grid overlays. Reeeived from https://upload.wikimedia.org/wikipedia/ commons/f/f9/Britain-fractal-coastline-50km.png, attributed to Wikipedia Commons, accessed 11/18/20.

57. *Fractal Diagnosis of Advancing Lung Cancer.* Progressive images of lung cancer fractally measured. Received from https://www.ncbi.nlm.nih.gov/core/lw/2.0/html/tileshop_pmc/tileshop_pmc_inline.html?title=Click%20on%20image%20to%20zoom&p=PMC3&id=4989864_nihms808299f6.jpg, attributed to usa.gov, NCBI, NLM, NIH, and attributed to Frances E Lennon 1, Gianguido C Cianci 2, Nicole A Cipriani 3, Thomas A Hensing 4, Hannah J Zhang 5, Chin-Tu Chen 5, Septimiu D Murgu 6, Everett E Vokes 1, Michael W Vannier 5, Ravi Salgia Lung cancer – a fractal viewpoint, grant support P30 CA014599/CA/NCI NIH HHS/United States, accessed 11/19/20.

58. *Butterfly Effect.* After Lorenz, these figures show two segments of the three-dimensional evolution of two trajectories (one in blue, and the other in yellow) for the same period of time in the Lorenz attractor starting at two initial points that differ by only $10-5$ in the x-coordinate. Initially, the two trajectories seem coincident, as indicated by the small difference between the z coordinate of the blue and yellow trajectories, but for t > 23 the difference is as large as the value of the trajectory. The final position of the cones indicates that the two trajectories are no longer coincident at t = 30. Retrieved from https://upload.wikimedia.org/wikipedia/commons/thumb/4/44/TwoLorenzOrbits.jpg/300px-TwoLorenzOrbits.jpg, attributed to Wikipedia Commons Butterfly Effect, accessed 11/19/20.

59. *Rossler Effect:* Three dimensional trajectories derived from Rossler differential equations that are non-coincident everywhere. Retrieved from https://upload.wikimedia.org/wikipedia/commons/thumb/7/7b/RosslerAttractor3D.svg/220pxRosslerAttractor3D.svg.png, Wikipedia Rossler effect, accessed 11/19/20.

60. *Julia Set and Mandelbrot Diagram.* Recursive pattern of one of Julia's patterns. The Mandelbrot set demonstrates the self-similar patterns of fractal geometry. Received from https://upload.wikimedia.org/wikipedia/commons/1/11/Julia _set_camp4_hi_rez.png, and received from https://upload. wikimedia.org/wikipedia/commons/a/a7/Mandelbrot_set_ima ge.png, and retrieved from https://upload.wikimedia.org/ wikipedia/commons/b/b3/Mandel_zoom_07_satellite.jpg att- ributed to Wikimedia Commons, accessed 11/19/20.

61. *Sharkovsky Real Line.* Given the Sharkovsky Theorem, the algorithm for deciding where the bifurcations will happened in a given period coincides with the intersection of a 45 degree line called the "real line". Original diagram.

62. *Bifurcation Map.* A chart of the logistics equation showing the blank areas where one or more bifurcations happen. Retrieved from https://upload.wikimedia.org/wikipedia/commons/ thumb/7/7d/LogisticMap_BifurcationDiagram.png/800px- LogisticMap_BifurcationDiagram.png , attributed to Wiki- media Commons public domain, accessed 11/28/20.

63. *Telegraph Circuit Schematic.* A possible circuit schematic seg- ment of a telegraph line. Original illustration.

64. *Particle Entropy.* As an entity dies or loses its potential energy, intropy increases the the particles separate indicating a lower energy state. Original diagram.

65. *Heat Transfer.* Educational poster of heat transfer with ther- mometers. Retrieved by https://thumbs.dreamstime.com/ z/heat-transfer-physics-poster-vector-illustration-diagram-heat- bal-ancing-stages-heat-transfer-physics-poster-vector-116756068. jpg, attributed to and purchased from dreamstime.com, accessed 11/16/20.

66. *Normal Distribution.* Gaussian distribution or Bell curve. Retrieved from https://upload.wikimedia.org/wikipedia/commons/thumb/7/74/Normal_Distribution_PDF.svg/1200px-Normal_Distribution_PDF.svg.png, attributed to Wikipedia Commons, accessed 11/19/20.

67. *Maxwell-Boltzmann Distribution.* A distribution resulting from the collision of quantum particles. Retrieved from https://upload.wikimedia.org/wikipedia/commons/1/19/Maxwell-Boltzmann_distribution_pdf.svg, attributed to Wikipedia Commons, accessed 11/19/20.

68. *Maxwell-Boltzmann Speed Averages.* Because the Maxwell-Boltzmann distribution is not symmetrical, the most probable speed does not correspond to the midpoint or the average velocity of a Normal Distribution with velocity. Original diagram.

69. *Reverse Entropy Distribution.* As particle temperature/speed increases, the Maxwell-Boltzmann distribution extends and the energy of ionization separation becomes probable. Retrieved as https://upload.wikimedia.org/wikipedia/ commons/thumb/1/19/Maxwell-Boltzmann_distribution_pdf. svg/ 800px-Maxwell-Boltzmann_distribution_pdf.svg.png, att-ributed to Wikimedia Commons, accessed 11/28/20.

70. *Battle of Britain Memorial – St. Paul's Cathedral Still Stands.* St Paul`s Cathedral on the Battle of Britain Memorial, on the Victoria Embankment - the symbol of resistance during the Blitz having remained standing while all around was demolished. Retrieved from https://thumbs.dreamstime.com/z/london-uk-july-st-paul-s-cathedral-battle-britain-memorial-victoria-embankment-symbol-resistance-blitz-having-1679 79951.jpg, attributed to and purchased from dreamstime.com, accessed 11/16/20.

71. *Bandpass Filter Circuitry for Speakers.* A passive bandpass filter circuit with output curve generally used in speaker design to separate the frequency spectrum for different size speaker cones. Original drawing plus illustrations retrieved from https://thumbs.dreamstime.com/z/speakers-13709972.jpg, and retrieved from https://upload.wikimedia.org/wikipedia/commons/1/19/Maxwell-Boltzmann_distribution_pdf.svg, attributed to and purchased from dreamstime.com and attributed to Wikipedia Commons, accessed 11/20/20.

72. *Bandpass Filter of Stock Noise.* A bandpass filter using mathematics outlined by John Ehlers reveals a waveform that ignores the jitter of a stock curve. Original chart.

73. *Stringed Instrument Waveforms.* Given a length of string for a violin or other instrument, where you place your finger determines the waveform and, therefore, the pitch of the musical note. Retrieved from https://upload.wikimedia.org/wikipedia/commons/thumb/2/2f/Moodswingerscale.svg/800px-Moodswingerscale.svg.png, attributed to Wikipedia Commons, accessed 11/28/20.

74. *40 Minute Chart (208 Hours, 312 Bars).* Tuning a bandpass filter waveform to stock curve peaks, the near-term period of peak stock action can be established. Original chart.

75. *20 Minute Chart (avg.: 88 Hours, 265 Bars).* By freezing the bandpass filter parameters from the tuned 40 minute chart, peak stock action on a faster chart with 1 to 2 scaling can be revealed. Original chart.

76. *13 Minute Chart (avg.: 78.5 Hours, 361 Bars).* Scaling down 1 to 3 from the 40 minute chart, more detailed peak action can be revealed. Original chart.

77. *10 Minute Chart (avg.: 46 Hours, 277 Bars).* The final 1 to 4 scaling from the 40 minute chart reveals precise peak detail on the 10 minute chart. Original chart.

78. *Nile River Mean Time for Poor Water Conditions.* The bandpass filter reveals a waveform frequency that can be converted to time; in this case, the average number of years between poor water conditions on the Nile River. Original chart.

79. *Newton's Cradle.* Transfer of energy through hanging balls. Retrieved from https://thumbs.dreamstime.com/z/newton-s-cradle-demonstration-conservation-momentum-energy-wrecking-balls-isolated-white-background-newton-s-cradle-201381347.jpg, attributed to and purchased from dreamstime.com, accessed 11/15/20.

80. *Electron Verses Earth Mass.* Illustration of atom structure including electrons alongside of a NASA photo of the planet Earth to contrast their sizes. Retrieved from https://thumbs.dreamstime.com/z/simple-model-atom-structure-electrons-orbiting-nucleus-three-protons-neutrons-science-kids-cartoon-style-vector-126940251.jpg, and retrieved from https://www.google.com/url?sa=i&url=https%3A%2F%2Fwww.nasa.gov%2Ftopics%2Fearth%2Findex.html&psig=AOvVaw1Pmec3P7bakwGLuhQ0O-jd&ust=1605974364836000&source=images&cd=vfe&ved=0CAIQjRxqFwoTCODG1be_ke0CFQAAAAAdAAAAABAv, attributed to and purchased from dreamstime.com and attributed to NASA, accessed 11/20/20.

81. *Electron Internuclear Distance.* Bonding graph of electron distance from nuclear attraction. Original illustration.

82. *The Shape of Electron Orbits.* Electrons orbiting at different energy levels resulting in three dimension illustrations of orbital directions. Retrieved from https://upload.wikimedia.org/ wikipedia/commons/thumb/4/4a/Single_electron_orbitals.jpg /250px-Single_electron_orbitals.jpg, attributed to Wiki-media Commons, accessed 11/20/20.

83. *Quantum Electron Orbits.* De Broglie suggested that electrons orbiting around the nucleus could be acting as waves, forming circular standing waves based on the size of the orbit. Just as a string which could only support a discrete set of wavelengths between its fixed nodes, an integer number of electron waves would have to fit within an orbital. Received from https://www.ck12.org/flx/show/THUMB_POSTCARD/imag e/user%3AY2sxMnNjaWVuY2VAY2sxMi5vcmc./98045-13700 42952-45-91-electron-standing-waves.png, attributed to

Accessed 11/29/20.

84. *Ionization Energy Levels for Electron Separation.* Plots of the energy ionization levels for row 3 chemicals in the periodic table. Original chart.

85. *Energy Ionization of Iron (Fe).* Chart connecting the six orbital ionization levels for iron. Original diagram.

86. *Pythagorean Tetractys.* Triangle shaped formation for numbers 1 through 10. Original illustration.

87. *Plato's Lambda.* A triangle of numbers formed by multiplying by 2 down one side and 3 on the other. Original illustration.

88. *Hieroglyphic Template.* A template to created original hierogly-phics seems to be a bed of 64 squares that encompass a circle and six triangles. Original illustration.

89. *Babylonian Numbers.* The ancient way of marking numbers in clay. Retrieved from https://upload.wikimedia.org/ wiki-pedia/commons/thumb/d/d6/Babylonian_numerals.svg/1024 px-Babylonian_numerals.svg.png   By Josell7 - File:Babylon-ian_numerals.jpg, CC BY- SA 4.0, https://commons. wiki-media.org/w/index.php?curid=9862983, attributed to Wiki-media Commons, accessed 11/20/20.

90. *Hieroglyphic Words.* Words in various hieroglyphic and ancient languages. Retrieved from https://upload.wikimedia.org/ wikipedia/commons/thumb/f/f9/Cuneiform_evolution_from _archaic_script.jpg/290px-Cuneiform_evolution_from_archaic _script.jpg, attributed to Wikipedia Commons, accessed 11/20/20.

91. *Sumerian Addition.* Each digit is read from right to left and is root 60 per digit.. Original illustration.

92. *Squaring the Circle.* In order to discover the area of a circle and the value of pi, you needed an inner and outer polygon whose area was known. The technique was called squaring the circle and is one of the longest running mathematical problems in history. Original illustration.

93. *Archimedes Pi.* Stick man standing next to blank white board explaining and pointing. Retrieved from https:// thumbs. dreamstime.com/z/stick-figure-explaining-pointing-blank-board-vector-illustration-man-standing-next-to-white-39394866 .jpg, attributed to and purchased from dreamstime.com, accessed 11/16/20.

94. *Euclid's Approximation of Pi.*   By using a six sided octagon, Euclid's geometry made it clear that a triangle angle of 60 degrees meant that pi must be greater than 3. Original illustration.

95. *Knife Triangles.* An illustration of the "knife arrangement" by its author Michael Schneider, The Golden Ratio at the Dinner Table, that shows a progressive connection with $\sqrt{3},\sqrt{(5,}$ and $\varnothing$). Original illustration.

96. *Dimensional Proportion Venn Diagram.* A two term proport-ion means that common values are shared as pictured in a Venn diagram. Original diagram.

97. *Koch Snowflake Potential Energy Growth.* Fractal geometry exercise using triangles that progressively divides into smaller triangles while adding area to a square side. Retrieved from https://thumbs.dreamstime.com/z/koch-snowflake-construct-ion-fractal-geometry-exercise-using-triangles-progressively-div-ides-smaller-white-black-160870568.jpg, attributed to and purchased from dreamstime.com, accessed 11/16/20.

98. *Kinetic and Potential Energy.* Physics forces visualization with bicycle and mountain. Retrieved from https://thumbs.dreamstime.com/z/kinetic-potential-energy-explanation-label ed-vector-illustration-scheme-physics-forces-visualization-bicyc-le-mountain-184852957.jpg, attributed to and purchased from dreamstime.com, accessed 11/16/20.

99. *Kepler Telescope Journey to Find Life.* Kepler Telescope's search for planets. Retrieved from https://www.nasa.gov/sites/default/files/styles/full_width/public/thumbnails/image/kepl er_eof_08-01-ws-oil-paint_filter_0.jpg?itok=3CVbInvU, attrib-uted to NASA, accessed 11/24/20.

100. *Fractal Nature*. The Grand Canyon of Romania is a protected area, a natural monument of national interest in Alba County, Romania. It is a geological and botanical reserve, located in the extreme southwest of the Secaşelor Plateau on the right bank of the Seca ul Mare, about 4 kilometer north of Sebe . Retrieved from https://thumbs.dreamstime.com/z/slope-pink-rock-cliffs-green-trees-rapa-rosie-grand-canyon-romania-under-clear-blue-sky-r%C3%A3%C2%A2pa-ro%C3%A8%E2%84%A2ie-rromania-124867750.jpg, attributed to and purchased from dreamstime.com, accessed 11/24/20.

101. *Aswan High Dam*. Built between 1960 and 1970, the Aswan High Dam, is the world's largest embankment dam built across the Nile in Aswan, Egypt. Retrieved from https://thumbs.dreamstime.com/z/aswan-high-dam-aswan-egypt-89942182.jpg, attributed to and purchased from dreams time.com, accessed 11/21/80.

102. *Consciousness*. A simple graph showing the evolution of group conscience. Retrieved from https://upload.wikimedia.org/wikipedia/commons/thumb/2/2f/Moodswingerscale.svg/1200px-Moodswingerscale.svg.png, attributed to Wikipedia Commons, accessed 11/20/20.

103. *Expanding Squares*. Radiating, expanding squares. Retrieved from https://thumbs.dreamstime.com/z/radiating-expanding-squares-geometric-monochrome-black-wh-white-element-royalty-free-vector-illustration-81771250.jpg, attributed to and purchased from dreamstime.com, accessed 11/16/20.

104. *Kronecker Product Sierpinski Triangle*. This internet on-line Kronecker product came from a 2017 internet article entitled Generating Kronecker Product Based Fractals, Dr. A. E. Voevudko, architect. Original illustration.

105. Kronecker Product Fractal Examples. More Kronecker Product examples of fractal images generated from an internet on-line program based on a 2017 internet article entitled Generating Kronecker Product Based Fractals, Dr. A. E. Voevudko, architect. Original illustrations.

106. *Cat's Cradle String Game.* Ancient children's game played internationally with approximately 40 inches of string. Retrieved from https://upload.wikimedia.org/wikipedia/commons/thumb/d/de/Cats-cradle.svg/1024px-Cats-cradle.svg.png, and retrieved from https://upload.wikimedia.org/wikipedia/commons/6/62/%D4%B9%D5%A5%D5%AC%D5%A1%D5%AD%D5%A1%D5%B2_6.png, and retrieved from https://upload.wikimedia.org/wikipedia/commons/6/68/%D4%B9%D5%A5%D5%AC%D5%A1%D5%AD%D5%A1%D5%B2_11.png, and retrieved from https://upload.wikimedia.org/wikipedia/commons/3/33/%D4%B9%D5%A5%D5%AC%D5%A1%D5%AD%D5%A1%D5%B2_13.png, and retrieved from https://upload.wikimedia.org/wikipedia/commons/a/ac/%D4%B9%D5%A5%D5%AC%D5%A1%D5%AD%D5%A1%D5%B2_15.png, and retrieved from https://upload.wikimedia.org/wikipedia/commons/3/32/%D4%B9%D5%A5%D5%AC%D5%A1%D5%AD%D5%A1%D5%B2_18.png attributed to Wikimedia Commons in the Public Domain, and attributed to Alfred Cort Haddon or Stephen Chapman Simms or William Henry Furness III - Jayne's String Figures and How to Make Them (1906), p. 327 – 336 and attributed to Wikipedia Commons in the Public Domain, accessed 11/24/20.

Bibliography

1. Anderson Arthur, *Introduction to the Transmission Line*, All about Circuits, Nov. 12, 2015.
2. Blanchard Paul, Devaney Robert, Hall Glen, <u>Differential Equations</u>, Third Edition, Brooks/Cole, Cengage Learning, Belmont, 2006.
3. Brunes Tons, <u>The Secrets of Ancient Geometry – and its use</u>, Vol. 1 & 2, Rhodos, Copenhagen, 1967.
4. Calder Nigel, <u>Einstein's Universe</u>, Penguin Books, New York, 1986.
5. Campbell David, *Nonlinear Science from Paradigms to Practicalities*, Los Alamos Science Special Issue, 1987.
6. Choi Charles, *The Universe's Dark Ages: How Our Cosmos Survived*, Space, Oct. 24, 2011.
7. Chou Felicia, Hawkes Alison, Cofield Calla, *NASA Retires Kepler Space Telescope, Passes Planet-Hunting Torch*, NASA news release, Oct. 30, 2018.
8. Courteau Stephane, Woodrow Lawrence, McDonald Michael, Guhathakurta Puragra, Gilbert Karoline, Zhu Yucong, Beaton Rachal, Majewski Stephen, *The Luminosity Profile and Structural Parameters of the Andromeda Galaxy*, Astrophysics Journal, June 17, 2011.
9. Cross Michael, *Brownian Motion*, California Institute of Technology: Physics 127b: Statistical Mechanics, Oct. 6, 2006.
10. Ehlers John, *A Peek into the Future*, metasoftware.com, accessed Dec. 2, 2020.
11. European Southern Observatory, *The odd couple: Two very different gas clouds in the galaxy next door*, phys.org, Aug. 7, 2013.

12. Gazale Midhat, <u>Gnomon from Pharaohs to Fractals,</u> Princeton University Press, Princeton, 1999.

13. Grime James, *A meandering tale, the truth about pi and rivers*, Guardian News and Media Limited, 2020.

14. Halpern Joshua, 10.5: *Covalent Bonding: Lewis Structure*, LibreTexts Chemestry, July 1, 2019.

15. Hawking Stephen, <u>A Brief History of Time</u>, Bantam Books, New York, 1988.

16. Heisenberg Werner, <u>Physics and Beyond: Encounters and Conversations</u>, Harper Touchbooks, translated from German by Arnold Pornerens, New York, 1972.

17. Henderson Tom, *Kepler's Three Laws*, The Physics Classroom: Physics Tutorial: Circular Motion and Satellite Motion: Lesson 4, 1996-2020.

18. Hoadley Renee editor, *Article 132: The Holographic Universe - Part 1 - Fractals & Holography*, Cosmic Core.org, Oct. 2018.

19. Kung David, <u>How Music and Mathematics Relate</u>, The Great Courses, Chantilly, 2013.

20. Labade Rekha and KshitijaSanap, *Design and Analysis of Compact Fractal Patch antenna with Enhanced Bandwidth and Harmonic Suppression for WiMAX Applications*, International Journal for Research in Engineering Application & Management (IJREAM) Special Issue – ICRTET-2018 ISSN : 2454-9150, 2018.

21. LaViolette Paul, <u>Earth under Fire</u>, Starburst Publications, Schenectady, 1991.

22. Lennon Frances, Cianc Gianguido, Cipriani Nicole, Hensing Thomas, Zhang Hannah, Chen Chin-Tu, Murgu Septimiu, Vokes Everett, Vannier Michael, and Salgia Ravi, *Lung cancer—a fractal viewpoint, HHS Public Access*, July 14, 2015.

23. Lundy Miranda, Sutton David, Ashton Anthony, Martineau Jason, <u>Quadrivium</u>, Walker & Company, New York, 2010.

24. Makeig Scott, *Means, Meaning, and Music: Pythagoras, Archytas, and Plato*, University pf California San Diego: Dept. of Music, 1981.

25. Mandelbrot Benoit, The (Mis)Behavior of Markets, Profile Books, London, 2008.

26. Marett Doug, *A Replication of the Silvertooth Experiment*, Conspiracy of Light, 2012.

27. Nave R, *Maxwell Speed Distribution Directly from Boltzmann Distribution*, Hyperphysics: Heat and Thermodynamics, accessed Dec. 12, 2016.

28. Ohlef Henry, The Second Judgment, Temporal Publishing, 2017.

29. Oldershaw Robert, *A Fractal Geometry?*, Amherst College Geology Dept., Amherst, 2003.

30. Peitgen Heinz-Otto, Jurgens Harmut, Saupe Dietmar, Chaos and Fractals New Frontiers of Science, Springer-Verlag New York,New York, 1992.

31. Piscarev Dimitriy, *Calculating the Hurst Exponent*, MetaQuotes Metatrader 5, Apr. 3, 2017.

32. Plato, Timeaus and Critias, Penguin Books, London, 1977.

33. Ruelle David, Chance and Chaos, Princton University Press, Princton, 1991.

34. Senovilla Jose and Garfinkle David, *The 1965 Penrose singularity theorem*, Classical and Quantum Gravity, Oct. 2014.

35. Stone Mark, *The Cubit: A History and Measurement Commentary*, Journal of Anthropology Article 489757, 2014.

36. Strogatz Steven, Nonlinear Dynamics and Chaos, Perseus Books Publishing, Cambridge, 2000.

37. Sutcliffe John, Hurst Stephen, Awadallah Ayman, Brown Emma & Hamed Khaled, *Harold Edwin Hurst: the Nile and Egypt, past and future*, Hydrological Sciences Journal Volume 61 - Issue 9: Facets of Uncertainty, 2016.

38. Talbot Michael, <u>The Holographic Universe</u>, Harper-Perennial, New York, 1991.

39. Tiller William, <u>Science and Human Transformation</u>, Pavior Publishing, Walnut Creek, 1997.

40. University of Liverpool, *Part 3: The Maxwell-Boltzmann Gas*, PHYS393 – Statistical Physics, accessed Oct. 7, 2020.

41. UNSW School of Physics, *Lecture 7: Maxwell Boltzman & Particle Motion*, Physics 2060, Sydney, 2009.

42. Utrecht University Faculty of Science, *Physicists build fractal shape out of electrons*, Physics.com news, Nov. 12, 2018.

43. Walters Patrick and Hey Tony, <u>The Quantum Universe</u>, Cambridge University Press, New York, 1994.

44. Wikipedia, *Godel metric*, Wikipedia encyclopedia, updated July 12, 2020.

45. Wikipedia, *Kirkwood Gap*, Wikipedia encyclopedia, updated July 14, 2020.

46. Wikipedia, *Olbers' Paradox*, Wikipedia encyclopedia, edited July 20, 2020.

47. Wikipedia, *Penrose diagram*, Wikipedia encyclopedia, edited Oct. 13, 2020.

48. Wikipedia, *Random walk*, Wikipedia encyclopedia, edited Sep. 26, 2020.

Index

Meet the Author

Henry Ohlef has authored five books: three are science fiction novels and two are on the piano works of classical composers Beethoven and Liszt. He worked for 31 years as a principal engineer for Allied-Signal Aerospace, taught microcircuit verification, and wrote and edited numerous technical manuals. He graduated with a Bachelor's Degree from Michigan State University and a Master's Degree from the University of Illinois in electrical engineering. Currently retired and living in South Florida, he enjoys flying, golf, and symphony orchestras.

The book titles follow:

- Arthur's Spirit
- Circles of the Suns
- The Second Judgment
- An Amateur's Hypothesis for Listening to Beethoven's 32 Piano Sonatas
- 105 Abstracts on the Piano Music of Franz Liszt